CoCo

The Colorful History of Tandy's Underdog Computer

CoCo

The Colorful History of Tandy's Underdog Computer

BOISY G. PITRE • BILL LOGUIDICE

CRC Press
Taylor & Francis Group
Boca Raton   London   New York

CRC Press is an imprint of the
Taylor & Francis Group, an **informa** business

CRC Press
Taylor & Francis Group
6000 Broken Sound Parkway NW, Suite 300
Boca Raton, FL 33487-2742

© 2014 by Taylor & Francis Group, LLC
CRC Press is an imprint of Taylor & Francis Group, an Informa business

Printed on acid-free paper
Version Date: 20131023

International Standard Book Number-13: 978-1-4665-9247-6 (Paperback)

Library of Congress Cataloging-in-Publication Data

Pitre, Boisy G.
    CoCo : the colorful history of Tandy's underdog computer / authors, Boisy G Pitre, Bill Loguidice.
        pages cm
    Includes bibliographical references and index.
    ISBN 978-1-4665-9247-6 (pbk.)
    1. TRS-80 computers--History. 2. Tandy Corporation--History. I. Loguidice, Bill. II. Title.

QA76.8.T18P58 2013
621.39--dc23   2013040642

Visit the Taylor & Francis Web site at
http://www.taylorandfrancis.com

and the CRC Press Web site at
http://www.crcpress.com

Contents

Preface

This is a story of a special computer. It is a computer that, over its life, became much more than just the sum of its plastic, metal, and silicon parts. Although it wasn't the most popular or most recognized system of its day, it was arguably the most loved. It had numerous magazines and conferences devoted to its existence, and earned the fondness of a multitude of fans. So strong was that regard that its adherents affectionately gave it the moniker "CoCo." Of course, I'm talking about the Color Computer, the 1980s home computer sold through the massive merchandising arm of Tandy Corporation and its thousands of Radio Shack stores and franchises across the United States, Canada, Australia, and Europe.

The CoCo would go through three major incarnations through its roughly 10-year retail product life: the original battleship gray TRS-80 Color Computer, the near-white Radio Shack Color Computer 2, and the final—and most advanced—Tandy Color Computer 3. During its time in the spotlight, the Color Computer taught kids and grown-ups alike how to program, provided students with the ability to quickly finish homework and write term papers, ran payroll for small businesses, made minifortunes for software and magazine publishers, and inspired young people to pursue careers in art, education, computer science, and engineering, to name just a few.

My own career choice, and by extension my path in life, was predicated entirely on my first encounter with a Color Computer. The year was 1983, and my childhood friend, Timothy Johnson, inherited a battleship gray Color Computer from his older brother Scott, who had taken his then-brand new Color Computer 2 to Louisiana Tech University to use in his studies. My fascination with this machine led to my spending many weekends at the Johnson house, exploring the BASIC manual and learning the fundamentals of bits, bytes, and programming. As nightfall approached, the only light that could be seen at the end of the hall was the green glow of the television under the bedroom door; the solitary sound was the cryptic droning of the cassette player loading up the adventure game

Raaka-Tu, the arcade clone *Donkey King*, or a BASIC program. Before I realized it, I had spent the entire night typing in programs or playing games, with the rays of an early morning sunrise piercing through the window as a brilliant reminder.

Two years later, I would spend the entire summer of 1985 mowing neighbors' lawns in order to save up enough money to buy a Tandy Color Computer 2 with 16K of RAM and Extended Color BASIC. With a computer of my very own, I stayed up many more late nights. My parents, witnessing my enthusiasm and realizing it wasn't a passing fad, supported my endeavor with additional hardware: an FD-500 floppy disk system for Christmas of that year and a 128K CoCo 3 with a CM-8 RGB monitor a year later.

All through my high school years, the CoCo 3 and a copy of the latest issue of *THE RAINBOW* magazine were my constant companions. I was introduced to the OS-9 Level Two operating system through a distant CoCo club, and began to learn 6809 assembly language. It was at that point that I decided to pursue a career in computer science and went to college. In 1992, I attended my very first CoCoFEST, then shortly thereafter moved to the Midwest to work for Microware Systems Corporation, the makers of OS-9. The rest, as they say, is history.

For the past several years I entertained the idea of writing a book on the history of the Color Computer. After some initial investigation and research, I began writing in earnest, then invited noted gaming author Bill Loguidice to come on board to assist me as coauthor. Bill graciously accepted, and his experience in both writing and navigating the publishing trails has been invaluable.

I would be remiss if I did not acknowledge some of the people who personally supported my endeavors along the way. First and foremost, I owe eternal gratitude to my parents for their support over the years, as well as my friend Tim Johnson for showing me what a CoCo was; Microware's Mike Burgher, James Jones, Ken Kaplan, Kim Kempf, and others with whom I worked; friend and business partner at Cloud-9, Mark Marlette; and the formidable CoCo community are all owed my appreciation. Last, but certainly not least, I must thank my wife, Toni, who has shown patience and understanding, both with my hobbies and with the many nights that I spent interviewing people and working on this book.

My Color Computer journey has been a wonderful one, and this book is just another milestone in that journey. Now you get the chance to read the fascinating inside stories, meet the interesting people, and learn just how this amazing computer came to be. It is my sincere hope that you enjoy the book.

—Boisy G. Pitre

*   *   *

It's a given that all computer technology, no matter how advanced at the time, eventually becomes obsolete. Although we may love the specific smartphone, tablet, videogame console, desktop, laptop, or other computing platform and associated operating system we're using at the moment, it's a fact that even the greatest technology will eventually become passé or simply wear out its welcome. Once that happens, most of us simply leave the past behind and move on to the next greatest technological wonder for another finite chunk of time before the whole process repeats.

What may not be such a given is the fact that when certain computer technology eventually gets ignored by the masses and heads to that big technology graveyard in the sky, that's not always the end of the story. Sometimes, when a platform has that certain something special, its most dedicated fans give it a second chance at life. Sure, this second chance may not get it shelf space at the local megastore, but it does ensure that a small but dedicated community of enthusiasts can continue to enjoy it. The Color Computer is one of those platforms.

Certainly, in comparison to classic Apple, Commodore, and Atari platforms, the CoCo is not particularly well known or remembered today. This is nothing new for the CoCo, however, as, for the most part, despite being showcased in the omnipresent Radio Shack catalogs and stores, the platform didn't get nearly the respect it deserved during its first 10 years of life, either. As such, the "Underdog" in this book's subtitle was no trivial addition.

In fact, I bet that for many readers, without that subtitle, you might have been just as likely to have thought that "CoCo" stood for the legendary French fashion designer, or, for the bad spellers out there, a dramatic retelling of the history of the cocoa bean. Luckily, that subtitle is there, so it's a good bet you knew exactly what you were getting into even if you never heard of the Color Computer. What you might not have known though in comparison to some of those other classic platforms is that even if the CoCo community's numbers are not quite as expansive, their support for their chosen platform is no less enthusiastic.

It's this very support that has made this book possible. From the original Tandy executives and engineers, right through to today's active product creators and information keepers, the CoCo's story has been one interesting, unbroken chain of individuals dedicated to the platform's past, present, and future.

The CoCo really is the ultimate underdog home computer, so it was without hesitation that I agreed to help document its fascinating story with one of its most enthusiastic fans and supporters, Boisy Pitre. As a passionate collector of vintage technology myself, I'm grateful for the opportunity to become a bigger part of the CoCo's amazing community. It's my wish that through this very book you'll get to experience some of what all the fuss was and is still about, and that you too will want to join in on the fun.

—Bill Loguidice

Acknowledgments

This book would simply have not been possible without the help of many people. From executives and engineers who were directly involved in the Color Computer's creation, all the way to the grassroots community of folks who wrote software, developed hardware, or just enjoyed the computer, their contributions were critical to the creation of this work.

We were fortunate to be assisted by many former Tandy/Radio Shack employees who took time to speak with us. Radio Shack President Bernie Appel and former Tandy CEO John V. Roach gave us the "big picture," while former directors Dr. John Patterson and Van Chandler filled in the details. Former engineering manager Chris Kline provided some background on the early years, as did engineers Ellis Easley, Walter Parkerson, and Paul Schreiber. Dennis Tanner provided us key information on Tandy's educational endeavors with the CoCo, and Mark Yamagata was instrumental in giving us the marketing perspective. Gary V. Pack provided insights into Tandy's legal maneuverings at the time. We would especially like to thank engineer Jerry J. Heep, and modern-day RadioShack Corporation's Eric Bruner and Paige Guyton for giving us access to engineering design notes, internal memos, and other files for reference while writing the book. And last, but certainly not least, we owe a great deal of thanks to the Color Computer's "evangelist," Mark Siegel, who provided copious amounts of material and checked and rechecked our work for accuracy. The "fathers" of the Color Computer also deserve special recognition: engineers Dale Chatham and John Prickett were very generous with their time in answering questions, reading, rereading, and commenting on drafts of our work to make sure we got the stories straight.

If Tandy created the CoCo's body, Motorola provided its soul. We would like to recognize the contributions of former Motorola employees John Dumas, Mike Kabealo, Stan Katz, Bill Peterson, Terry Ritter, and Steve Tainsky, all of who made themselves available for interviews. Together, they painted a clear picture of the formative years of the Color Computer.

While not every CoCo owner used OS-9, they all owe a debt of gratitude to Microware Systems Corporation, whose operating system undoubtedly helped to extend the CoCo's life. We would like to thank Andy Ball, Larry Crane, and Ken Kaplan for providing key parts of the early OS-9 story as it related to the Color Computer, as well as general Microware history. Other former Microware employees who made time to help were Eric Crichlow, Peter Dibble, Tim Harris, Mark Hawkins, Allen Huffman, and Kim Kempf.

There were several magazines dedicated to the CoCo over the course of its life, but none more formidable and important than *THE RAINBOW*. The magazine's former staff was extremely helpful in providing information on the inner workings of this important periodical. We would like to acknowledge Fred Crawford, Dan Downard, Jutta Kapfhammer Helm, Tamara Dunn Jarvis, and Greg Law for their interviews. Special thanks go to Jim Reed, who spent a considerable amount of time discussing the early years of Falsoft, and for graciously providing suggestions for the chapter on *THE RAINBOW*.

We received significant help from other early CoCo notables as well. Rick Adams, William Barden, Jr., and Dale Lear all graciously provided time for interviews, as did Dennis Bathory-Kitsz, Chris Burke, Tom Dykema, David Figge, Arthur Flexser, Al Hartman, Frank Hogg, Rich Hogg, Bruce Isted, Mike Knudsen, Ron Lammardo, Dale Puckett, Dave Stampe, Tom Roginski, Logan Ward, and Greg Zumwalt.

The members of the modern CoCo community gave us enormous encouragement and help. Daniel Campos and Juan Castro provided information on Brazil's Color Computer scene, as did Glen Van Den Bigelaar for Canada, Nick Marentes for Australia, Dejan Ristanovic for Yugoslavia, and Torsten Dittel for most of Europe. Dean Leiber's work in compiling numerous magazines, manuals, online messages, and other materials in electronic form assisted our research greatly. Mike Haaland, Tim Lindner, John Linville, Mark Marlette, Mike Pepe, Mike Rowen, Tom Seagrove, Eddy Szczerbinski, and Aaron Wolfe also provided information and support. Michael D'Alessio at www.RadioShackCatalogs.com readily provided us with high-resolution scans, for which we are also appreciative.

Finally, we want to thank the wonderful CoCo community, whose enthusiastic and loyal members all played their part in this tribute to Tandy's most beloved computer.

Long live the CoCo!

Authors

Boisy G. Pitre has been an avid and passionate advocate for the Color Computer since his early teens. In 1992, he joined Microware Systems Corporation, the makers of OS-9, as a software engineer, and has worked in the industry ever since. He remains a member of the CoCo community, leading various open source initiatives and working with Cloud-9 to provide hardware and software for CoCo hobbyists. Boisy is a graduate of the University of Louisiana at Lafayette where he obtained his master's and bachelor of science in computer science with a minor in mathematics. He and his wife, Toni, reside in the quiet countryside of Prairie Ronde, Louisiana.

Bill Loguidice is a critically acclaimed technology author and documentary producer, as well as cofounder and managing director for the online publication *Armchair Arcade*. A noted videogame and computer historian and subject matter expert, Loguidice personally owns and maintains well over 400 different systems from the 1970s to the present day, including a large volume of associated materials. He resides in Burlington, New Jersey, with his wife and regular coauthor, Christina, and his two daughters, Amelie and Olivia.

Tandy Gets Personal

The genesis of the Color Computer, or CoCo, as it would come to be known colloquially, began in 1919 in an unlikely manner with ancient technology: the tanning of hides and skins of animals. Shortly after the end of the First World War, two friends, Norton Hinckley and Dave L. Tandy, started the aptly named Hinckley-Tandy Leather Company, which supplied leather shoe parts and supplies to shoe repair shops in the Fort Worth, Texas, area.

In 1921, another dynamic duo, brothers Theodore and Milton Deutschmann, added some high-tech flavor to the CoCo's origin story by opening the first Radio Shack in Boston. Borrowing the name for the store from the literal name for the original wooden room used for housing radio equipment aboard U.S. Navy ships, Radio Shack similarly specialized in supplying ship radio equipment and "ham" radios. It was in supplying these amateur radio enthusiasts, or hams, who utilized the latest commercially available radio technology for noncommercial, mostly entertainment-related communications purposes, that either company would first directly target the consumer market. In many ways, targeting hams in the 1920s would parallel to targeting computer hobbyists in the 1970s, with both leading to bigger and better things to come in the subsequent decades.

Shortly after entering the burgeoning hi-fi music market in 1939, Radio Shack issued the first of its famous series of catalogs (Figure 1.1). In 1947, Radio Shack opened its first audio showroom, which featured speakers, amplifiers, turntables,

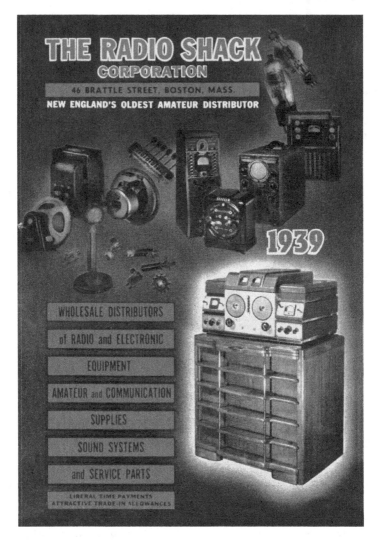

Figure 1.1

The Radio Shack Corporation's first catalog appeared in 1939 and would continue to be mailed out for over 50 years. (Courtesy of www.RadioShackCatalogs.com.)

and phonograph cartridges. By 1954, Radio Shack began selling its own products in its stores and catalogs under the brand name Realist, but, due to a legal dispute, settled on Realistic, which lasted as a brand until officially being discontinued in 2000.

While Radio Shack was first expanding its operations, Hinckley-Tandy Leather Company was going through its own series of changes. Civilian leather rationing early on in World War II devastated the business, forcing a lateral move to support an emerging leathercrafting hobby, where the company could get a supply priority by providing for the armed forces. These changes were implemented by Dave Tandy, but inspired by his son, Charles D. Tandy, who graduated from Texas Christian University and attended Harvard Business School before joining the U.S. Navy for the remainder of the war. While enlisted, Charles Tandy saw the value in leathercraft as a therapeutic tool and recreational activity for patients at military hospitals and believed it could carry the company through to postwar success. This vision, combined with some of his other practical achievements, which included setting a record for selling war bonds, predicted his future success running the family business.

Charles Tandy returned to Fort Worth in 1947 to join the family business for good. At that time, the Hinckley-Tandy Leather Company was a five-store operation with a thriving mail-order catalog and sales of about $750,000 per year. Despite the obvious success, Tandy, driven by the same entrepreneurial spirit that ignited way back in his childhood and never stopped burning, wanted more.

The young Tandy's vision for a leather-clad future that included profound support for hobbyists and nationwide dominance clashed with Hinckley's own desires for the company. As a result, the Hinckley-Tandy Leather Company officially split on March 31, 1950, with the two Tandy's forming the Tandy Leather Company and Hinckley staying within his comfort zone and keeping the shoe findings business.

By 1954, the Tandy Leather Company grew to 67 independently operated stores in 36 states and the Hawaii Territory, with sales of $8 million annually. Unfortunately, it soon became impractical to keep expanding while remaining a privately held family business with several dozen employee-shareholders. As a result, in October 1955, the Tandy Leather Company was sold to the struggling American Hide and Leather Company of Boston, whose main appeal was a listing on the New York Stock Exchange. The new entity changed its postmerger name in December 1956 to General American Industries.

Following a string of unsuccessful acquisitions that stymied the very growth Charles Tandy was after, the ever-ambitious businessman began what would become a four-year struggle to gain control of the merged company. In November 1959, Tandy succeeded, being elected board chairman and chief executive officer. The following year he moved the corporate headquarters back to the familiar confines of Fort Worth and changed the company's name to Tandy Corporation. By November 1963, the company's stock began trading on the New York Stock Exchange under the symbol TAN, where it remained until May 2000 when the company name was changed to RadioShack Corporation and a listing symbol of RSH.

Meanwhile, by the early 1960s, Radio Shack, specializing in electronic parts and related hobbyist products, had expanded to nine retail outlets plus a mail-order business. Unfortunately, a disastrous decision to sell product on eventually uncollected credit brought the company to its financial knees, and, with no shrewd leadership in place to save the business, all seemed lost. Enter Charles Tandy, who saw the potential in consumer electronics and the possibilities for rapid growth that the Radio Shack chain offered. So, in November 1963, for what equated to $300,000, or about $2.1 million adjusted for inflation, Radio Shack came under the Tandy Corporation umbrella, which at the time consisted of more than 170 leather and leathercraft stores in the United States and Canada, with annual sales of more than $20 million. This bargain of a transaction set in motion a massive American success story that would eventually see Tandy phase out its nonelectronic product line completely by 1975, spinning off its vestigial enterprises into Tandycrafts and Tandy Brands. In what can only be considered fortuitous timing, 1975 was also the year that the Mits Altair 8800 microcomputer was introduced to the world, with Tandy Corporation now in a prime position to take advantage of this new type of product: the personal computer. To see how these two events came into alignment, however, we must first return to World War II.

While Charles Tandy's involvement in World War II was pivotal for the future direction of the company, so too were what would be born from the war's computational demands, which coincidentally led to an industry that would also come to change the world. In December 1943, British engineer Tommy Flowers and his team secretly unveiled the world's first electronic, digital, programmable computer, the Colossus Mark 1. Operational by early February of the following year, the Colossus and its successors were used by British codebreakers to read encrypted German messages.* These codebreakers included the likes of Alan Turing,† whose seminal 1936 paper on the notion of a "universal machine" capable of performing the tasks of any other machine demonstrated that anything computable could be represented by 1s and 0s, which proved critical when trying to decipher codes with 15 million million possibilities. The improved Colossus Mark 2 went into operation on June 1, 1944, in time for the pivotal Normandy landings. In total, 10 Colossus computers were placed into operation by war's end. Unfortunately, their strict focus on code breaking and overall secret nature—which was kept into the 1970s—limited the potential of their influence.

Like the armed forces for the other world powers at the time, the U.S. Army was on a continuous quest to gain an upper hand against its enemies, and several promising—and some not so promising—projects were given funding on the off chance that a few might be successful. One such proposal was to create a high-speed electronic device to calculate ballistics firing tables, which at the time was being performed manually by female "computers," or literally "one who computes."

As a result of that proposal, development of the Electronic Numerical Integrator and Computer—better known as ENIAC—began on June 5, 1943; however, it did not become fully operational until 1946 when it became the first reprogrammable, electronic general-purpose computer. Conceived and designed by John Mauchly and John Eckert, the room-sized ENIAC was a modular

---

* A large number of electromechanical calculating devices known as "bombe" helped decipher encrypted messages specifically from the infamous German Enigma machines.

† Turing contributed several pivotal mathematical and computing ideas in his tragically short life. For instance, in a famous 1950 paper he posed the thought-provoking question, "Can machines think?" which led to the famous "Turing test" concept, which is a measure of a machine's ability to exhibit indistinguishably human behavior.

computer, composed of individual panels that performed different functions. More flexible than the Colossus and not constrained by the secrecy of the war effort, the ENIAC was able to more profoundly influence the development of later, increasingly smaller and more powerful computers from a variety of commercial companies. Thus began the transition from centuries-old mechanical and analog paradigms to the purely digital.

The bulky and unreliable vacuum tubes used into the 1950s were phased out by more reliable and less expensive transistors in the 1960s. These transistors were soon incorporated into the integrated circuit (IC), where a large number of these semiconductor devices were placed onto small silicon chips. Nevertheless, after several decades of innovation in circuitry and refinements in operation and utility—including a switch to a stored-program methodology that offered a fully reprogrammable environment—large and expensive mainframe computers still remained the norm.

Despite size and cost restrictions that limited these computing systems to government and large institutions such as universities, games found their way onto even the earliest mainframes and represented an important step in making these finicky behemoths more approachable. The Nimrod, a single-purpose computer designed to demonstrate the principles of digital computing to the general public by playing the game of Nim, was showcased at the Exhibition of Science during the 1951 Festival of Britain. Despite having an abstract grid of light bulbs that passed for its display, it was nevertheless an important milestone. Interestingly, in that same year and into the next, the United Kingdom played host to other innovations, with the Pilot ACE computer simulating draughts (checkers), and the Ferranti Mark 1 computer playing host to the first computer-generated music, as well as one of the first attempts at solving chess problems.

One of the earliest recognizable computer games was Alexander Douglas's 1952 creation of *OXO*, a simple graphical single player versus the computer tic-tac-toe game on the Electronic Delay Storage Automatic Computer (EDSAC) mainframe at the University of Cambridge. Although more proof of a concept than a compelling gameplay experience, *OXO* nevertheless set the precedent of using a computer to create an immediately accessible virtual representation of a real-world activity.

In 1958, for a visitors' day at the Brookhaven National Laboratory in Upton, New York, William Higinbotham and Robert Dvorak created *Tennis for Two*, a small analog computer game that used an oscilloscope for its display. *Tennis for Two* rendered a moving ball in a simplified side view of a tennis court. Each player could rotate a knob to change the angle of the ball, and the press of a button sent the ball toward the opposite side of the court. As with *OXO*, few people got to experience *Tennis for Two*, but in many ways it can be considered the first dedicated videogame system, which themselves are just simplified personal computers. Without the benefit of hindsight, though, this milestone was even lost on the game's creators, who, after a second visitors' day one year later, disassembled the machine's components for use in other projects. Historically tragic to be sure, but intrinsic of the hacker mindset of use and reuse that would come to guide the personal computer revolution in the years to come.

It wouldn't be until 1962 that the most influential early computer game, *Spacewar!*, blasted onto the scene. Initially designed by Steve Russell, Martin Graetz, and Wayne Wiitanen, with later contributions from Alan Kotok, Dan Edwards, and Peter Samson, the game was the result of brilliant engineering and hundreds of hours of hard work. Developed on the DEC PDP-1 mainframe at MIT, *Spacewar!*'s gameplay was surprisingly sophisticated and ambitious, pitting two spaceships against each other in an armed duel around a star that exhibited gravitational effects on the two craft. Each player controlled a ship via the mainframe's front-panel test switches or optional external control boxes, adjusting each respective craft's rotation, thrust, fire, and hyperspace, which is a random, evasive screen jump that may cause the user's ship to explode. Over the years, the game was improved many times and inspired many clones and spiritual successors, including the first commercially sold arcade videogame in 1971, *Computer Space*, which was designed by Nolan Bushnell and Ted Dabney, who would go on to found Atari just one year later.

It was this ability to inspire that was perhaps *Spacewar!*'s greatest contribution to the future of computing. Even with still privileged access to the host hardware limiting the game's wider exposure, enough of the personal computer industry's[*] key future movers and shakers got to see firsthand that these machines could be used for something more than serving the often sober computational needs of businesses, universities, and the government. In short, they could also delight and entertain the individual.

With the needs of the individual in mind, Drs. Thomas Kurtz and John Kemeny took the next major step toward computing for the masses at Dartmouth University with the creation of the Beginner's All-purpose Symbolic Instruction Code, or BASIC, programming language, in 1964. While low-level programming languages, like machine code or assembly language, provided little or no abstraction from a computer's native instructions and thus ran more efficiently than a high-level language like BASIC that needed to be complied or interpreted before execution ever could, they also required strong mathematical and scientific skills to have a fighting chance to understand, let alone use effectively. With BASIC, greater emphasis was instead placed on natural language syntax and simple logic, something that enthusiasts of all ages and disciplines could appreciate and more easily work with. In another stroke of genius, Kurtz and Kemeny made their BASIC compiler available free of charge in hopes that the language would become widespread. This altruistic strategy worked, with variations of their original BASIC language becoming a staple on several key computing systems of the time, before going on to dominate the first few decades of personal computing, setting the standard for how most users would first learn to program, including on the CoCo.

---

[*] And it was becoming a "computer" industry, as the terms "electronic brain" or "mechanical brain" were slowly being phased out in popular usage.

Unfortunately, despite being shown a clear path to wider computer acceptance, including with the likes of Douglas Engelbart's "The Mother of All Demos," in late 1968, which featured the first coordinated demonstration of key components of future connected personal computers, including the mouse, hyperlinks, and video conferencing; Alan Kay's "Dynabook" concept (1968–1972), which predicted form factors and use cases realized by today's laptops and tablets; and, starting with its founding in 1970, Xerox PARC's stunning working-office environment that featured locally networked desktop computers with bitmapped graphics, graphical user interfaces, object-oriented programming, and "what you see is what you get" (WYSIWYG) output to high-quality laser printers, a major roadblock for the industry at large remained into the early 1970s.

That roadblock was the main operating model of the time, which was to leverage a single large computer that needed to be shared and have its time partitioned among many users. This model was certainly effective and a significant improvement over previous decades when a single-user's activity would tie up a computer for hours, or even days, but it did not scale well, and proved costly for the would-be end user to access. A change was needed. That change would come from Intel.

By late 1971, Intel had developed and released the first mainstream microprocessor, or single-chip central processing unit (CPU), the Intel 4004, which became the heart of many small-scale digital computer projects. Without the microprocessor, accompanying refinements, and constant improvements, there would be no home computer market, since hybrid analog and digital technology suffered from substantial limitations in size and versatility (Figure 1.2).

Although microprocessors held great promise for home computer applications, it took another three years before they started to catch on with manufacturers, and then, begin to impact consumers. Thus, important technological innovations and games continued to appear almost exclusively on mainframe systems throughout the 1970s. Many of these games, like the early dungeon-crawling game, *dnd* (1974), by Gary Whisenhunt and Ray Wood for the versatile PLATO computer instruction mainframe system, and Will Crowther's PDP-10 computer game *Adventure* (1975)—one of the first text adventures—inspired entire genres when home computers finally rose to prominence in the mainstream consumer marketplace. This rise began in late 1974 with the development of the aforementioned MITS Altair 8800 computer kit, based on the Intel 8080 microprocessor released earlier in the year. Advertised in the January 1975 edition of *Popular Electronics* magazine, the kit was an unexpected success, enthusiastically supported by groups of eager hobbyists who had long waited to get their hands on a computer they could call their own (Figure 1.3).

The Altair 8800 had a red LED display and several toggle switches to directly program the system. There were no other display or input options, and little could be done with the default configuration, but the system was a step up from prior kits and plans that required the potential user to track down their own parts. Most of the Altair 8800's intelligence was built using removable cards, making the motherboard—the heart of the computer that handles system resources—a means to interconnect the components. Since the motherboard accepted 100-pin

Figure 1.2

As the cover of this 1972 catalog demonstrates, by the 1970s, Radio Shack, now a major part of Tandy Corporation, featured an impressive range of electronics. Although personal computers would still be several years away, Radio Shack was already well positioned for the coming revolution. (Courtesy of www. RadioShackCatalogs.com.)

Figure 1.3

The cover of the legendary January 1975 issue of *Popular Electronics,* which kicked off the personal computer revolution. Art Salsberg's page 4 editorial boldly proclaimed, "The Home Computer Is Here!" Ironically, the original Altair computer was lost in transit, so the system featured on the cover was just a rough, nonworking mockup.

expansion cards, it eventually became known as the S-100 bus, which was an important industry standard into the early- to mid-1980s, and was often found in computers paired with the versatile CP/M operating system. In fact, by late 1975, the first of many greatly improved clones appeared in the form of IMS's IMSAI 8080 kit, which itself ran a modified version of CP/M, and was made famous in 1983 as one of the "stars" of the movie *War Games.*

In spite of, or perhaps because of, mainframe resources being such a precious commodity, working with these large machines was rarely a solo pursuit. Small but dedicated and enthusiastic communities gathered around these fascinating technologies, with the free flow and sharing of knowledge embodying the hippy ideal of the late 1960s. This sense of community translated to the home market at the earliest opportunity, with the Altair 8800 providing the first real platform to rally around. As a result, computer hobbyist user groups started to spring up, with the most famous being the Homebrew Computer Club in Silicon Valley, which first met in March 1975. The club's early appeal was the enthusiastic and free exchange of ideas, technology, and information among its talented members. Club membership consisted not only of hobbyists, but also engineers and other professionals, like future Apple cofounders Steve Jobs and Steve Wozniak, who, among other notables, would go on to shape the path of the computer industry for the next several decades.

Interestingly, a young and brash Bill Gates, then of Micro-Soft, wrote an open letter for the Homebrew Computer Club's second newsletter condemning commercial software piracy. His Altair BASIC, cowritten with Paul Allen, was the first language available for the system, and hobbyists were illegally copying the desirable, but expensive, paper tape software in droves. This letter marked the first notable rift between the ideals of free software development and the potential of a nascent retail software market. Gates and the quickly-renamed Microsoft went on to create versions of BASIC, operating systems, and other types of software for nearly every personal computer, including most of Tandy's, creating an important, if infamous, business empire in the process.

With computer user groups and clubs on the rise, specialized retail stores starting to meet some of the demand for product, and publishers churning out in-depth enthusiast magazines like *Byte*, general interest in computers was quickly spreading. The first major computer fair, the Trenton Computer Festival, took place in April 1976 and featured speakers, forums, user group meetings, exhibitor areas, and an outdoor flea market, setting the template for future trade shows. As computers made the transition from countless kit offerings to complete, prebuilt systems in 1977, they became more appealing to a wider segment of the population, much like the Trojan horse for the core technology, the home videogame console, which was conceived by engineer Ralph Baer as far back as the 1950s.

Unfortunately for Baer and those who might have hoped for a quicker personal computing revolution, even his concept for a television videogame was so novel that he was unable to garner enough support to build working prototypes until the mid-1960s. His first attempt to build a home videogame console was a simple game of tag featuring two squares ("Chase"), which soon morphed into his famous "Brown Box" prototype. The prototype included several additional diversions, including a paddle and ball, and target-shooting games. After getting rejected by several TV manufacturers, Baer finally signed an agreement in 1971 with Magnavox, which released a refined version of the prototype the following year, renaming it the Odyssey Home Entertainment System.

Perhaps the Odyssey's most enduring legacy was inspiring Nolan Bushnell at a Magnavox product demonstration in 1972. Later that same year, Bushnell founded Atari and, with engineer Al Alcorn, developed *Pong* for the arcade. *Pong* had one simple directive, "Avoid missing ball for high score." A smash hit was born.

Pong itself was clearly a derivative of one of the Odyssey's paddle and ball ("Tennis") games, and Atari's success with *Pong* led several other companies to also copy the game's concept. Magnavox eventually filed a successful lawsuit against Atari for copyright infringement, forcing the fledgling company to settle for a lump sum and other manufacturers to pay hefty licensing fees.

Pong's simple but compelling gameplay was in stark contrast to Bushnell's and Dabney's earlier *Computer Space* by Nutting Associates. Despite its striking cabinet design, relatively large and inviting screen, and four straightforward action buttons for fire missile, thrust, rotate left, and rotate right, *Computer Space* was too complex for the general public's first exposure to videogames. Bushnell later admitted that the game appealed mostly to his engineering friends who had enjoyed its inspiration, *Spacewar!*. Though less impressive in nearly every way than *Computer Space*, it was *Pong*'s emphasis on approachability and fast-paced fun that would first define and then establish an entire industry. In each of their own way, these object lessons in accessibility would be the hallmark of all the great computing successes in the years to come.

Although the Odyssey received a small sales boost from the popularity of *Pong* and the various clones that sprung up in the arcade, the console never achieved critical mass popularity in American homes, selling a few hundred thousand units before it was discontinued in 1975. This was due in part to the limited marketing and distribution afforded through Magnavox retail outlets, something that a chain like Radio Shack, with its national marketing campaigns, catalogs, and many more store locations, would have been able to mitigate. The fact that the Odyssey came from Magnavox also gave less informed consumers the impression that it would only work on Magnavox televisions. This same type of inextricable brand and product association would initially help Tandy and Radio Shack in gaining consumer trust (after all, there was little chance of Radio Shack going anywhere) but would later prove something of an albatross when it became difficult to shake off inevitable missteps in quality control or product releases over the years.

When Atari created a home version of *Pong* in 1975—complete with automatic scoring and sound—the dominant retailer at the time, Sears, agreed to distribute it under its own brand name, Tele-Games, to great success, legitimizing the viability of Baer's plan to market videogame systems for home use. Atari released its own branded version of the console starting in 1976, just as an explosion of *Pong* clones saturated the home videogame market. Although these machines were extremely popular for a time, and offered increasingly sophisticated feature sets, there were simply too many systems for the market to sustain them all for long. This was particularly the case in light of the rise of fully programmable consoles that used interchangeable cartridges for more diverse gameplay possibilities, starting with Fairchild's Video Entertainment System (VES) in 1976. This home videogame breakthrough was followed one year later on the home computer side with the release of the preassembled and relatively user-friendly Apple II, Commodore PET, and Tandy's own TRS-80 systems, each of which featured its own interchangeable software, first on cassette tapes and, later, disks. This legendary trinity marked the first time fully assembled, programmable computers were readily available to and usable by the masses.

With changes to a few key decisions here or there, any of the trinity could have made it to market first, but that honor ultimately fell to the Apple II, which was formally introduced at the West Coast Computer Faire on April 16, 1977, and first sold on June 4. The starting price for a basic unit with 4K of RAM was $1,298 (about $4,600, adjusted).

The tale of the Apple II begins with the two Steves from Sunnyvale, California: Steve "Woz" Wozniak, a talented engineer specializing in calculators at Hewlett-Packard (HP), and Steve Jobs, who was an energetic and eccentric summer employee there. Woz had been friends with Jobs in high school, where the two hackers had made money selling "blue boxes," illegal devices used by "phreakers" (phone system hackers) to steal free long-distance and eavesdrop on private conversations.

Jobs became Atari's 40th employee in 1974, serving the innovative young company as an hourly technician. He left Atari for a yearlong hiatus to India, returning to work with a shaved head and traditional Indian garb. Atari had scored big with its arcade version of *Pong* and was about to repeat its success with the play-at-home version. Jobs, now a night-shift engineer thanks in part to his eccentricities and inability to relate to many of his coworkers, was asked to create a prototype for a single-player, vertical *Pong* variant called *Breakout*, whose goal was to clear rows of blocks at the top of the screen by bouncing a ball off a small, movable paddle at the bottom. Unfortunately, the technology required to create a *Breakout* machine would tear into its profits, so Atari wanted a design that used as few chips as possible. Faced with such a daunting engineering challenge, Jobs sought the help of his old friend, Woz, something that company management was hoping for anyway.

Atari had witnessed Woz's impressive self-built home *Pong* clone but had failed to woo him away from HP. Nevertheless, Woz, a fan of both Atari arcade games and engineering challenges, came to his friend's rescue. He completed the bulk of the work in about four days, with an efficient design that used far fewer chips than any other Atari arcade game at the time. Atari's engineers were impressed and Jobs received a nice payout and bonus—most of which he famously kept for himself. *Breakout* would become another arcade hit for Atari.

After years of hardware hacking and his two dalliances in videogames, Woz began work on a television computer terminal. Woz realized that one major stumbling block for the nascent home computer industry was the lack of a cheap and effective means of displaying output. Computer hobbyists could either content themselves with a row of flashing LEDs, or ante up for an expensive video display or clunky text terminal. None of the available options favored anyone but the most dedicated of hobbyists.

Inspired by the creative and highly motivated group at the Homebrew Computer Club, Woz was soon demonstrating a prototype that would ultimately become the Apple I, which was known simply as the Apple Computer prior to the release of its successor. Really nothing more than an elegantly designed circuit board with a low-cost MOS 6502 microprocessor, 4K RAM, and expansion connectors, the two Steves' first computer nevertheless laid the critical foundation for what was to come. Atari and HP were not interested in the prototype, and negotiations with a calculator company, Commodore, fell apart, so the duo formed their own company, Apple Computer, on April 1, 1976.

Working out of Woz's bedroom and Jobs's garage, the two soon began production on the Apple Computer. The legendary Jobs's persuasiveness (aka "Reality Distortion Field") was in full force even then, and he negotiated with a local hobbyist computer store, the Byte Shop, for an order worth $50,000 (about $190,000 today). Credit, time, and supply constraints were tight, but the Byte Shop order was met, with the computer store providing full-stroke keyboards and wooden cases to complement the circuit board. Through the Byte Shop and magazine coverage and advertisements, the fledgling company had appreciable, if slow growth, from Apple computer sales.

Even before the first Apple computer had been officially released, Jobs and Woz were already thinking up new features; they frequently updated the design and shared their progress with the Homebrew Computer Club. The result was the Apple II. Even though little time had passed since their first release, the new unit improved on the Apple I in nearly every way. It sported a complete molded plastic enclosure with a full-stroke keyboard, external peripheral ports, and eight easily accessible internal expansion slots.

Woz, who enjoyed dazzling his friends at the club, wanted to play a version of *Breakout* written entirely in BASIC. Such a feat would have been unthinkable on the Apple I, so Woz's design for the Apple II came to incorporate color graphics commands, circuitry for paddle controllers, and a speaker for sound. With these standard features in place, the Apple II offered technology that its rivals in 1977, the Commodore PET and Tandy TRS-80, could not match. In fact, it would take Tandy until 1980, with the Color Computer, and Commodore, in 1981 with the VIC-20, before either company would even make an attempt. Of course, there were other ways both Commodore and Tandy were competing with the Apple II, including price and distribution, making the early victor in the home computer wars anything but a foregone conclusion.

Although the Commodore PET was the first of the trinity to be introduced, in this case at the Winter Consumer Electronics Show (CES) in January 1977, the first few hundred units would not ship until October. The PET's origins were with Chuck Peddle and a small team of engineers who worked at Motorola in 1973 on the powerful but expensive 6800 microprocessor. Peddle and other members of this team left Motorola soon after to join MOS Technology, where the features of the Motorola 6800 were mirrored and improved upon but at a significantly reduced cost with a chip called the 6501. To avoid legal issues with Motorola, the chip received a slight redesign and was reintroduced as the 6502, which would eventually appear in one form or another in a large number of videogame and computer systems well into the late 1980s.

Commodore Business Machines (CBM) used MOS Technology chips in their popular line of electronic calculators, but major parts supplier Texas Instruments decided to enter the lucrative chip market in 1975. This market shift caused supply and cash flow problems for companies such as Commodore and MOS. With a cash infusion from the main investor and chairman Irving Gould, Commodore was able to purchase several suppliers, including MOS Technology, to assist in the company's goal of reducing dependence on outside manufacturing. Jack Tramiel, founder and president of the renamed Commodore International, Ltd., stipulated that as part of the acquisition, Peddle join the company as head of engineering.

In this position, Peddle helped to convince Tramiel to change the focus of the company from calculators to computers. Peddle's previous work on MOS Technology's successful single-board kit computer, the 6502-based KIM-1 (which Commodore would later come to rebrand for a time), inspired the basic design of the Commodore PET 2001, or simply PET (named for the pet rock craze of the time*), which was the first complete, all-in-one home computer.

Interestingly, just prior to this pioneering period, there was the potential for close working relationships between Commodore, Tandy, and Apple. Since Commodore was having difficulty fully funding the initial production of the PET, Tramiel approached Tandy with the idea of exclusively selling the computer. Though interested, Tandy nixed the idea when Tramiel, ever the aggressive businessman, insisted that the company must also buy a huge supply of calculators. Nevertheless, Commodore was still able to proceed on its own by preselling the PET several months before it was actually ready to ship, hence the large gap between its initial announcement and the first units reaching consumers. At Tandy, they simply went ahead with their own computer design. As for Apple, during development of the Apple II, Steve Jobs was looking for $300,000 in funding, while Tramiel and Commodore, who also supplied that system's 6502 processor chips, only offered $50,000. Jobs and Apple went with a different group of investors, ensuring that all three companies would be set to make their own mark in history.

Commodore's design origins showed in the first version of the PET 2001, which featured a metal chassis and keyboard with oversized, closely spaced, calculator-style keys, whose stiff plastic design and resemblance to the chewing gum gave it the nickname of "chiclet." While this unusual form of input was as much about leveraging existing product in Commodore's supply chain as anything else, Tramiel insisted that since the PET was a new type of device, there was no reason to utilize paradigms from typewriters. However, the chiclet design, copied in spirit in a variety of future computers from several manufacturers, did not allow for practical touch typing, frustrating countless users. On the flip side, the PET's keyboard did feature an additional numeric keypad, something rarely implemented even in competitive systems released much later and at higher cost. The machine's standard features were rounded out by the built-in cassette recorder and 9″ monochrome monitor, which handled the modest demands of

---

* Also known by the backronym of "Personal Electronic Transactor."

the PET's character-based graphics with aplomb, all for an initial starting price of $495 (about $1,760, adjusted) for the 4K RAM version.

Commodore, in a situation that would serve it well for many years to come, now controlled most of its own manufacturing and supply services, so it was able to provide relatively high-power systems at a lower cost than possible for most of their competitors. In addition, a version of the popular BASIC language from Microsoft, redubbed COMMODORE BASIC, was purchased for a one-time fee and licensed in such a way that it could be extended and reused without restriction in future Commodore machines with the same processor type, even within those from a different product line, such as the VIC-20, Commodore 64, and Plus/4. Microsoft would never allow this type of licensing again.

The last of the trinity's origins, the TRS-80, goes back to that fateful year in 1975, when, due in large part to the nationwide success of CB (citizen-band) radio sales, Tandy Corporation decided to focus exclusively on electronics. Unfortunately, the CB radio fad would soon begin to fade and Tandy would need a new direction to take them through the end of the decade. This new direction—personal computers—came from three key employees: John Roach, Don French, and Steve Leininger.

Roach, as vice president of manufacturing, and French, a buyer on the West Coast in Silicon Valley, were instrumental in bringing Leininger, who previously worked as an engineer at National Semiconductor, to Tandy, in late 1976. It was a fateful decision for the future of Tandy and the industry in general to sell a ready-to-go computer instead of a kit, which they all agreed was too difficult of a project. Thanks to Leininger's moonlighting at a computer store, he had particular insight into the challenges of the hobbyist market of the time.

"Some customers were having trouble assembling '100-in-1' [electronics] kits, let alone something as complex as a computer," recalled Leininger in an interview in the December 1981 issue of *Popular Computing*. After about six months on the job at Tandy and with the last of the internal hurdles overcome—deciding on the microprocessor type and the use of dynamic RAM—Leininger was able to officially begin work on a complete computer system.

The Tandy Radio Shack-80, branded as the TRS-80 (though sometimes described in contemporary references as "S-80" in an ultimately failed attempt to create a generic descriptor for the platform), was introduced on August 3, 1977, and released in September. Initial sales for the TRS-80 were strong, due in no small part to availability at all Radio Shack locations, which numbered over 3,000 stores at the time. Though its brief flirtation with selling a wide range of competitor computers, kits, and related accessories via a Tandy Computers catalog in 1978 was truer to the company's roots, they also knew a good thing when they had it and thereafter kept the focus firmly on the success of its own computers as long as they remained directly involved in the business (Figure 1.4). In fact, it was this company focused, regular, national advertising combined with the extensive reach of its Radio Shack stores that would prove a true competitive advantage for Tandy until other home computers became widely available from mass merchants in the early 1980s.

Figure 1.4

The back cover of the Tandy Computers 1978 catalog, showcasing the TRS-80. What makes this particular catalog special is that within its pages, Tandy also offered an extensive range of their competition's computers and accessories for sale and delivery—with the notable exception of Apple and Commodore products. The experiment was short-lived, however, as future catalogs would exclusively feature Tandy products. (Courtesy of Bill Richman at www.geektrap.com.)

As opposed to the MOS 6502 used in the Apple II and Commodore PET, the TRS-80 used the Zilog Z80 microprocessor. This low-cost chip, which provided the "80" in the TRS-80, proved a popular choice for many different manufacturers in future computers, including those based around the CP/M operating system, which would be the closest thing to an industry standard outside of the original trinity until the slow but eventually all encompassing rise of the IBM Personal Computer (PC) and, in particular, Microsoft's MS-DOS, starting in 1981.

The first TRS-80 featured a full-stroke keyboard within a thick base that housed all of its core components, a design emulated by many other manufacturers in the future, including Commodore, with its VIC-20. A numeric keypad did not become standard until later models in the series. The TRS-80 initially came with 4K RAM and an external 12″ black-and-white monitor, as well as a separate cassette drive, all for $599.95 (about $2,130, adjusted). Memory expansion was possible to 16K internally and more externally with the eventual release of an expansion interface box, which, among other things, allowed for the use of 5.25″ floppy disk drives.

Various software programs were targeted to one of the two versions of BASIC for the TRS-80. The simplified Level I BASIC, which was standard on the initial systems, fit into a 4K ROM chip and was based off a public domain version of the language. Level II BASIC was licensed from Microsoft and supported many more features but required a 12K ROM chip. Eventually, most TRS-80 systems would come with at least 16K of RAM and Level II BASIC, making it a popular standard to code to.

The TRS-80's operating system and external devices did not work smoothly at times, garnering the quirky computer the nickname of "Trash 80." Unfortunately for Tandy, even though overall quality was improved in future revisions and models, the cheeky nickname was hard to shake, and Trash 80 remains a popular designation for the generic TRS line of systems to this day.

The TRS-80 was renamed the TRS-80 Model I around the time of the 1979 release of the mostly incompatible business-centric Model II, which featured a high capacity 8″ disk drive integrated with the monitor. This would just be the beginning of Tandy's countless efforts to continuously expand its computer product lines.

Due to its inability to meet Federal Communications Commission (FCC) regulations regarding radio frequency (RF) device emissions, the Model I was discontinued in early 1981. Tandy, knowing this was coming, released the all-in-one Model III as the first model's true successor about a year earlier. Maintaining almost total compatibility and a disk conversion option, the Model III also added standard support for lowercase characters and featured a faster microprocessor. In 1983, the Model 4 was released as an upgrade to the Model III, and allowed the added option of running CP/M software.

Despite selling the most units and having the largest amount of available software of the trinity going into the end of the 1970s, Tandy was clearly starting to lose both mindshare and sales to a growing number of flashier rivals, particularly in the home, where built-in color graphics and sound were becoming an expectation rather than something nice to have. This new breed of relatively low cost but still highly capable personal computers was led by the 1979 introduction of Atari's 400 and 800, and Texas Instruments' TI-99/4. They certainly would not be the last. Clearly, something had to be done. Fortunately for Tandy, the seeds for its intended response in both target price and capabilities were already being sown.

2

Planting the Seed

By the late 1970s, having gone from a leather goods supply company to a retail electronics giant in the span of just a few decades, Tandy was already a king of unlikely origins and transformations. As such, it is hardly surprising that Tandy's first foray into color computing had an equally unlikely genesis, this time on the American farm.

Not long after Radio Shack made its splash in the home computer market with the TRS-80 Model I, Tandy officials were invited to take part in an exciting and ambitious project to bring computing technology to tobacco farmers in central and southwestern Kentucky. The year was 1978, and the project, known as the Farm Information Retrieval System (FIRS), was conceived as a way to electronically deliver data directly to where the farm owners lived and worked. Proposed in March of that year by Kentucky Senator Walter "Dee" Huddleston, the $380,000 initiative was sponsored by the U.S. Department of Agriculture (USDA) and the National Weather Service (NWS).

The resulting project was simply called "Green Thumb." The unusual name might have seemed inauspiciously similar to the title of Ian Fleming's James Bond-driven novel and film, *Goldfinger,* which was named for its lead villain's obsession with gold, but the project's moniker actually fit the term's origination perfectly. While today "having a green thumb" means someone who is skilled at gardening, the term actually came from colonial America, where farmers' thumbs would literally be stained green from cutting the stems of tobacco flowers with their thumbnails.

Green Thumb was envisioned as a farmer's "electronic almanac," where information such as commodity prices and weather data would be made available via a computer terminal at the farmer's location. This interface would provide the key information that the farmer could use to facilitate decision making on everything from crop management and cultivation practices to production costs and fertilizer and pesticide applications.

Operation was envisioned to be simple. The farmer would be presented with a menu of options, then would press a key corresponding to the information that he or she wished to see. The selection would then initiate a transfer of bits of information that traveled via a modulator/demodulator, or modem, and then over the venerable "Ma Bell" telephone system, whose twisted copper wire had covered a good part of the rural American landscape by the mid-1970s. The downloaded information could then be viewed and referenced as needed until another choice had been made or the terminal was powered down.

Expectations were high that Green Thumb would deliver: government officials estimated that the data provided by the project could contribute to a 25% reduction in pesticide use alone, making crop management cheaper and more cost effective. In order to facilitate its usage, the Green Thumb Box, as the terminal was to be called, would be offered free of charge to farmers. For the initial phase, 250 units would be made available in Kentucky: 100 would go to Shelby County, while another 100 would be sent to Todd County. The 50 additional remaining units would be kept for repair and testing purposes.

As Green Thumb began to gain momentum, the NWS sought bids from companies that could do both the design and manufacturing of the necessary hardware. Tandy Corporation was an obvious choice of bidder as a previous collaboration with the NWS led to the successful weather radio, then selling in thousands of Radio Shack stores across the country. Not only that, but Tandy's manufacturing capability in Fort Worth, Texas, was more than capable of handling the small number of terminals that were anticipated for the project.

When made aware of the project, Tandy management expressed interest, but there was a problem: the company was not set up to bid on government contracts. What Tandy needed was a partner that had experience drawing up bids for government-sponsored projects and could then be trusted to properly negotiate the ensuing contract. Tandy found that partner in Motorola. For years, Motorola had been supplying the government's space program with electronic parts and other products, and had razor-sharp experience in cutting through the bureaucratic red tape that so often accompanied the bidding process.

The synergy between the two companies was clear, so Motorola and Tandy drew up a technical and management proposal. The former would provide design and parts for the hardware, and the latter would lay out the printed circuit board (PCB), provide the casing, and then perform final assembly. Motorola's Stan Katz, who managed the systems engineering group and headed the strategic marketing group that would try to find markets for Motorola semiconductors, oversaw Joe Roy, who would manage the technical side of the project along with Steve Tainsky, William Peterson, and Jim Reid. Tandy's Director of Engineering, Dr. John Patterson, would take charge of the assembly and manufacturing process, with Chris Kline and Jerry Heep covering the engineering responsibilities. Motorola's Don Sheppard would coordinate from above as the project manager. From the government's end, USDA official Howard Lehnert would lead the project, while Dr. John Ragland of the University of Kentucky would provide the necessary academic support.

The joint design by Motorola and Tandy consisted of a single-board computer housed in a retrofitted TRS-80 Model I case, with a final design reminiscent of the era's television videogame consoles. Atop the case there would be no QWERTY keyboard. Instead, a 12-button keypad was located in the middle, resembling a touch-tone telephone with several extra keys on the right side. The brain of the terminal system would be the proven 3870 microprocessor, which was a variant of the Fairchild F8 CPU used in the first cartridge-based home videogame console, the Fairchild Video Entertainment System (VES), which was released in 1976.

Visual display of information received by the Green Thumb Box would be achieved by connecting the device to a standard television set via a built-in radio frequency, or RF, modulator (Figure 2.1). Much like a TV station broadcasting its programming, the RF modulator transmitted video on either channel 3 or 4, with the selection depending upon which received less interference in the user's area. The Motorola-designed 6847 Video Display Generator (VDG) integrated circuit, designed by Motorola's Peterson, was capable of generating 32 columns and 16 rows of text on the television, for a total of 512 viewable characters on a single screen. Each character would be comprised of an 8×12 dot matrix. Additionally, graphics capability would be available with up to eight colors in a

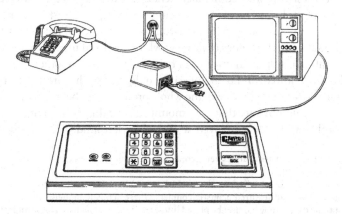

Figure 2.1

An illustration of the Green Thumb Box in operation. (From "The Green Thumb Proposal.")

64×32 coarse pixel resolution. A green-monochrome "full graphics mode" with a higher 128×96 resolution was also available.

Connectivity to the outside world was achieved by way of a built-in 300 baud* modem that connected the terminal to a telephone line,[†] which it shared with an existing phone by way of a splitter. To obtain information, the farmer simply picked up the phone and placed a call to a central computer, located somewhere within the county. After waiting for a few minutes for the data to download, the farmer would hang up the phone. Up to eight screenfuls, or pages, of information would then be available for off-line perusing, retained in the system's 4K or 8K of RAM as long as the unit stayed powered on.

As the project got underway, boards were soon produced and parts made available for assembly. For Jerry Heep, a former sergeant in the U.S. Marine Corps and an up-and-coming engineer who started working at Tandy in 1975, one aspect of the Green Thumb Box proved to be immensely challenging: the emission of radio frequency interference (RFI). RFI, a by-product of power being applied to electrical circuitry, can disturb or degrade the performance of onboard electric components or even those of other nearby electronics. Electrical engineers have wrestled with RFI for years in an attempt to manage it, since complete elimination was not possible. The Federal Communications Commission (FCC), the agency of the U.S. government charged with managing radio spectrum use, mandated limits on RFI in electrical products and forced electronic products to adhere to strict RFI guidelines in a section known as FCC Part 15.[‡]

It was not Heep's first run-in with RFI problems. The young engineer had wrestled with RFI before when designing an electronic game composed of a single integrated circuit. As Heep recalled, "You could hit that thing with a stick and the RFI would peak. It was totally unpredictable." Now, Heep was faced with trying to get the Green Thumb Box, with not one, but 28 integrated circuits, to pass the FCC's RFI tests. He warned management of the challenges that awaited them. Sure enough, try as he might, he simply could not bring the interference below the maximum allowable amount of 15 microvolts per meter. It did not matter that the terminal would be in the homes of farmers who often lived miles apart from their nearest neighbor, resulting in minimum risk of interference to others. The regulations had to be followed, and if engineering could not get the Green Thumb Box's electromagnetic interference down, the whole project could be doomed.

Once it became obvious that no amount of shielding was going to address the RFI problem, Tandy employed a clever workaround. "They issued me a

---

[*] Baud is equivalent to pulses or tones per second, so a 300 baud modem can deliver up to 300 bits of information per second.

[†] This is known as a direct connect modem, which was in stark contrast to most other modems of the time that instead featured acoustic couplers. These acoustic modems required that Bell System's standard telephone handset be placed into a cradle so it could listen for sounds that could then be converted to usable electrical signals.

[‡] It was a new revision of this same type of FCC regulation implemented on January 1, 1981, that caused Tandy to discontinue production of the noncompliant TRS-80 Model I in favor of the Model III.

transmitter's license, by the FCC, to operate an experimental television station," Heep later recalled. "The license would allow me to build transmitters which emitted a television signal." In essence, Heep's license allowed him to build and operate Green Thumb Boxes as individual little television stations. That stroke of end-run genius effectively got Tandy around the stringent FCC Part 15 regulations.

As Heep began to build the boxes, they were sent out into the field, distributed by county agents to farmers across the rural Kentucky countryside. Newspapers began to publish articles about Green Thumb, and awareness of the project started to build (Figure 2.2). Other parts of the country became interested in the promise of instant delivery of data to farms. Soon enough, over a dozen other states and their agricultural workers, including potato growers from as far away as Maine, wanted in on the action.

On March 3, 1980, after months of preparation, the switch was thrown and Project Green Thumb came online.

Project Green Thumb was deemed a success in meeting its objective as a useful tool for farmers. It was also a successful partnership between Motorola and Tandy, forging a working relationship in solving complex engineering problems that would continue into the coming decade. The project had also demonstrated that telecommunications was not only the way of the future, but with the right equipment and technology, something that was workable even under the most extreme and unusual circumstances.

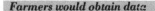

Farmers would obtain data

'Green thumb box' being developed

By DIANA TAYLOR
Associated Press Writer
LOUISVILLE, Ky. (AP) — Making hay while the sun shines, keeping cattle cool and setting tobacco before a spring downpour all may be made easier by a "green thumb box."

Such, at least, is the hope of developers of the gadget, researchers at the University of Kentucky's agricultural extension service.

A green thumb box basically resembles an electronic television game and works on a similar system, said Dr. John Ragland, an associate dean for the extension service.

The box would be plugged into a farmer's TV and telephone and connected on the other end with a centralized computer located somewhere in his county.

The farmer could dial the telephone number of the computer, activating his unit, and tell the computer what information he wants by using a keyboard on the green thumb box.

Then, the developers say, such data as current weather conditions, markets

and grain futures would appear on the television screen.

"By having a county computer terminal tied into our Lexington system — which is tied to the National Weather Service and the Chicago Board of Trade — we can) feed to the county computer this kind of information," Ragland said.

The green thumb box is a joint effort of UK and the U.S. Department of Agriculture. The idea originated when Kentucky's disaster warning system was in the planning stages, according to Ed Graves, a

DIANA
TAYLOR

Associated
Press
Writer

for other ways that we could provide weather information to people in Kentucky," Graves said in a telephone interview from his Washington office.

Housed tobacco

curing quickly

LOUISVILLE, Ky. (AP) — Many Kentucky farmers may be entertaining thoughts of preparing their burley for market now that much of the state's tobacco crop is in curing barns.

Harvesting conditions were almost ideal last week, and Kentucky farmers used

spokesman for Sen. Walter "Dee" Huddleston, D-Ky.

After the warning system was set up, "we were looking

the sunny days to harvest tobacco, corn, soybeans, hay and silage and to seed small grains, according to the Kentucky Crop & Livestock Reporting Service.

Although tobacco housing is 10 days to two weeks behind the normal schedule, the reporting service said 88

At that point, the National Weather Service pointed out that "if you had the right kind of facilities, you could localize weather information to the farmers."

The idea, which is new in the United States, will be tested in two Kentucky counties sometime after Christmas, Ragland said.

"It is more or less in operation in Great Britain and to a lesser extent perhaps in France," he said, "but we're taking quite a different approach."

The project is still in the planning stages, and the counties have yet to be chosen, Ragland said. "We'll draw up some procedures by which a county can apply for being chosen as one of two."

After the counties are

chosen, 200 farmers — 100 in each county — will be selected to participate.

The cost of the green thumb box is estimated at $100, but farmers who take part in the experiment will not have to pay for the unit, Ragland said. And there is hope that a successful experiment will lead to a lower price for the box.

There also is hope that success with providing farm information may lead to an expanded system of disseminating informaion, Ragland said.

"The idea, of course, is if the thing were to go, sooner or later a tremendous amount of information — not just weather and markets — would be available to a person with a system like this."

OPEN DAILY 9:30-9 SUNDAY 12-6

Figure 2.2

An article on the Green Thumb Box as it appeared in the *Park City Daily News*, October 2, 1978. (From Google Newspaper Archive.)

From Project Green Thumb, destiny had spoken. It was time to take the idea to the next level, in another project that was already underway at Tandy: the VIDEOTEX.

The early success with Project Green Thumb had reinforced the fantastic idea that the future promise of personal computing was remote communication, and that the ability to connect people with on-demand information was already technologically feasible and financially practical. For the Green Thumb user, its implementation meant giving the farmer critical schedules on crop rotation or fertilizer applications. But what about other information? What if the idea behind Green Thumb could be expanded to provide a broader range of information like news, stock quotes, sports scores, and weather to households all across America? What if that data could be accessed by anyone, at any time? A myriad of intriguing possibilities existed.

Something else became apparent during Project Green Thumb: the box had taken advantage of two specific items that were in roughly 95% of American households by the late 1970s: a telephone and television. Tandy, through its 5,000 Radio Shack stores—2,000 of which were located in small towns with populations of less than 15,000 people—had been selling accessories for both devices for years. As a result, Tandy knew its customers' demographics well.

It was not a stretch then for Tandy management to postulate a world where they paired the average consumer's telephone and television to a box that could communicate to the world. The possibilities were staggering, the potential market was huge, and Tandy was uniquely positioned to take advantage of the opportunity.

This vision of connectivity for the common man was none-too-soon in coming, for in the summer of 1979, CompuServe launched a groundbreaking online service called MicroNET. Up to that point, CompuServe's options for remotely accessed information services was limited to professional use, with the Columbus, Ohio, computer services organization supporting more than 650 commercial customers, including government agencies, financial institutions, and other large organizations. Through MicroNET, the average consumer microcomputer user was finally brought into the fold.

Positioned as a way for CompuServe to sell after-business-hours computer time for its mostly idle DECsystem-10 mainframe computers, availability was initially limited but relatively inexpensive for the time. For a one-time hookup charge of $9 and then just $5 per connect hour, home and small business users could access the computer time-sharing and software distribution service in 25 major metropolitan areas* between the hours of 6 P.M. and 5 A.M. on weekdays, and all day Saturday, Sunday, and holidays. While this may seem restrictive now, for the inquisitive personal computer user of the time armed with any capable terminal, access software, and a 300 baud modem, it was the only cost-effective

---

* MicroNET services were available in 153 other cities for an additional $4 per hour charge.

professional alternative to dialing into hobbyist Bulletin Board Systems, or BBSs.*
All that was needed now was for a company like Tandy to come up with a way for
the average consumer to more easily access the fledgling service.

It therefore appeared to be fortuitous timing that with Project Green Thumb's
launch, Tandy's mass-market-oriented successor, the TRS-80 VIDEOTEX
Terminal, was already under development. The VIDEOTEX name was not chosen lightly, as "Videotex" was the generic name at the time for one-way ("Teletext"
or, confusingly, "Videotext") or two-way ("viewdata") transmission using television screens as receivers. Videotex services were already rolled out to other parts
of the world, with each based on their own standards, and Tandy was striving to
be both literally and figuratively the name for such services in the United States.

Led by Heep, Tandy's VIDEOTEX, like Project Green Thumb, made extensive
use of Motorola parts, but this time placed them inside a gray case reminiscent
of the TRS-80 Model I's styling. A more versatile full 53-key chiclet QWERTY
keyboard replaced its predecessor's 12-button keypad. A DATA indicator light
was positioned on the top of the case, just above the keyboard to the upper right.

Internally, the VIDEOTEX consisted of three main components: the CPU, the
video processor, and the direct connect modem. The CPU contained the advanced
MC6809E 8-bit microprocessor, RAM, and supporting circuitry. The video processor was the same 6847 VDG integrated circuit used in Project Green Thumb, and
provided the same display resolution and graphics modes. Similarly, the internal
modem allowed for communications up to the same 300 baud speed.

Since its primary function was to provide access to remote online resources,
the VIDEOTEX was only capable of limited offline activities through the latter
two of its three operating modes: online, offline, and advanced. Online mode
was for communication with a host computer, while offline mode was for view-
ing data that was automatically saved when in online mode. Advanced mode
allowed offline data entry for sending when connected online to a host computer.
The base model came with 4K of RAM that could hold up to eight screenfuls,
or pages, of information, while expansion to 16K could save up to 32 screenfuls.

Despite its advanced internal architecture, the VIDEOTEX's external features, located on the rear of the unit, were minimal in deference to its intended
usage. There was a Reset button, a TV output (TO TV) to an RF antenna switch
box, a Channel Select (4 or 3) for the clearest television signal, a telephone port to
connect to a modular phone jack, and a Power button (Figure 2.3).

Unveiled on May 27, 1980, access to the VIDEOTEX was initially limited to
commercial enterprises and professional applications. Labels such as "AgVision,"
and partnerships such as the one with the Professional Farmers of America

---

* Telecomputing Corporation of America's The Source Information Utility was available around the
same time as MicroNET and was similarly featured. Although no less of an online pioneer, The
Source failed to achieve anywhere near the same subscriber numbers as CompuServe's service. A
higher initial start-up fee to subscribe and greater hourly charges likely contributed to its limited
success. In 1989, The Source was acquired by CompuServe and the service discontinued shortly
thereafter. CompuServe itself was acquired by AOL in 1998 and, though a shell of its former self,
continues to operate as an online services provider.

Figure 2.3

The TRS-80 VIDEOTEX Terminal as it appeared in the TRS-80 Computer Catalog, No. RSC-5, 1981. (Courtesy of www.RadioShackCatalogs.com.)

(PFA) to create the Instant Update database service, expanded upon Project Green Thumb's agricultural foundations and were used to brand VIDEOTEX Terminals. In some instances, these terminal cases were painted in a light blue color instead of the standard dull gray. The TRS-80 VIDEOTEX Terminal itself would be rolled out en masse to consumers in over 6,000 Radio Shack stores by November 30, 1980.

Tandy had no intention of leaving the burgeoning market of preexisting personal computer customers in the cold, either. VIDEOTEX-branded software packages started at just $19.95 (sans modem) and were released for everything from generic terminals to Tandy's own TRS-80 series computers. Even the competing Apple II got a version of VIDEOTEX.

While VIDEOTEX was designed to access its own custom services, BBSs, and the few commercial online services of the day like The Source and Dow Jones without bias, it was Tandy's exclusive two-year selling agreement[*] with CompuServe, which included Tandy-specific services, that was intended as its

---

[*] Michael A. Banks, *On the Way to the Web: The Secret History of the Internet and Its Founders* (Berkeley, CA: Apress, 2008), 61.

primary attraction. MicroNET soon began to be offered in Radio Shack stores, where many of the locations with computer centers were equipped with demo accounts to access the CompuServe Information Service.*

In turn, by the summer of 1980, MicroNET's availability at Radio Shack became a key component of CompuServe's own marketing, where the TRS-80 Model I, TRS-80 Model II, and VIDEOTEX were prominently featured in its advertisements (Figure 2.4). MicroNET and CompuServe Information Service users could gain access to all types of software, including networked multiplayer games, and a variety of personal, business, and educational programs, as well as 128K of online storage, bulletin boards for community affairs and for sale and wanted notices, electronic mail, and corporate stocks and public debt information. CompuServe promised even more was forthcoming to their rapidly expanding services menu, including regional newspapers, an electronic encyclopedia, travel information, and food preparation and gardening tips, all online features that, while strictly text-based, were still years ahead of their time.

Unfortunately for Tandy and their VIDEOTEX aspirations, while the idea that a large percentage of consumers would favor accessing online services over general computing activities would prove prophetic, it was approximately 15 years and one World Wide Web short of becoming reality. The fact that the VIDEOTEX Terminal was initially priced at $399 (about $1,100, adjusted), which made it comparable in price to a low-end personal computer, did not help its value proposition either. As such, the VIDEOTEX Terminal was a commercial failure. Even the competitively priced VIDEOTEX-branded software packages were soon surpassed in popularity by more versatile terminal software.

Although the VIDEOTEX experiment would prove a failure, luckily for Tandy, it was not its only new market strategy targeting personal computing at that time. Like many companies with talented leadership, a similarly big idea had already sprung up by the time the VIDEOTEX Terminal was under development. This formed a concurrent, two-pronged approach toward the rapidly evolving home computer market that ended up successfully acting as its own backup plan in case of one or the other's failure.

---

* In that same year, H&R Block would buy CompuServe to help develop electronic filings for tax returns. Although MicroNET was originally intended to represent the mainframe access services and CompuServe Information Service the information services, all such data offerings would soon fall solely under the latter name.

Figure 2.4

A CompuServe advertisement from the December 1980 issue of *BYTE*, featuring just three of Tandy's growing collection of platforms that could access the service.

3

Colorful Computing

As the 1970s were starting to wind down, it became apparent to Tandy Corporation management, including, most prominently, Radio Shack's new executive vice president John Roach,* that a new direction was needed to drive the company's personal computer ambitions in the coming decade. While the TRS-80 computer was still the single best-selling personal computer, with over 200,000 units sold by 1980 in comparison to just over 35,000 for the higher priced Apple II,† its 13% share of the midrange computer market was on a single-digit trajectory as an increasing number of high profile competitors, like Texas Instruments and Atari, entered the fray. Although Tandy Corporation would continue to manufacture and support its original monochrome TRS-80 line of computers well into the 1980s, staying competitive in the home consumer market would require some concessions to computer gamers—most notably with color and built-in support for sound, features that the Apple II and the majority of the newest competitors were already set up for.

Roach, who became Radio Shack's division president and chief operating officer in 1980, and, at age 42 in 1981 would become one of the youngest CEOs in the country, was one of the original forces behind Tandy's successful venture into preassembled computers. Clearly, he was no stranger to high-risk, high-reward

---

* Roach assumed the role not long after Charles Tandy's death at age 60 on November 4, 1978.
† Andrew Pollack, "Next, a Computer on Every Desk," *The New York Times*, August 23, 1981.

technology ventures. As such, it was decided that instead of trying to shoehorn competitive color graphics and sound capabilities into a new TRS-80, that computer line would remain separate and primarily be targeted to home office and business users (Figure 3.1). Fortunately for Tandy, in the interest of productizing

Figure 3.1

Tandy's marketing message for its TRS-80 line of computers would evolve over time to have a more professional, business-oriented focus, as this advertisement for the TRS-80 Model III from the July 1982 edition of *Popular Computing* magazine demonstrates.

the idea for a timely release in an increasingly competitive market, much of the groundwork for its first color computer was already laid by its earlier Project Green Thumb venture with Motorola Semiconductor, while other efficiencies could be leveraged with their concurrent VIDEOTEX initiative.

At roughly the same time Heep was doing his thing with the VIDEOTEX Terminal in his department, Dale Chatham was working on the new color computer in Tandy System Design. While each of the engineers worked separately, they both benefited from having each project share a majority of the necessary parts, with Motorola providing key support when needed.

According to Stan Katz, then head of Motorola's system engineering group, projects like this with Tandy came about because of the solid personal relationships between individuals at the two companies. Katz provided an example of the exceptional nature of this relationship by describing how he would personally pick up Tandy's John Patterson at the airport for meetings when they were both working out the logistics of Project Green Thumb. It was a given that Tandy was also a high priority for Katz's group because of the tremendous reach the thousands of Radio Shack stores provided for their semiconductors. It was no surprise then when Motorola produced a successful proposal in April 1979 for what would become the Color Computer.

Steve Tainsky, who worked as a systems engineering manager at Motorola and reported to Katz, led the group that came up with the Color Computer reference design. Tainsky's group did the architecture and logic design of the video display generator, synchronous address multiplexer, digital analog converter, and the supply and level translator chips. The silicon design for the chips was done by designers in product groups in Austin, Texas, and Mesa, Arizona. The peripheral interface adapter and microprocessor were already available, having been previously designed by a different Austin, Texas, internal group, led by Terry Ritter.

It was Tandy's Chatham who was just the guy to pull everything together and improvise the finished computer design. As Steve Leininger, who, after his pivotal role in the creation of the TRS-80 Model I, directed the development of the Model II, Model III, and Color Computer, recounted in the December 1981 issue of *Popular Computing*, "Dale was the right person to develop the Color Computer. He's a very good engineer. He's careful and does his planning the first time around."

Chatham's feelings for Leininger, who had resigned from Tandy shortly before publication of that issue of *Popular Computing*, were mutual. Chatham, who had earlier in his life studied electrical engineering (chosen over genetic engineering) in college and dropped out after a year when his 18-hour schedule plus part-time job proved excessive, dedicated himself to working the semiconductor processing line at Texas Instruments. The real-world Texas Instruments experience gave him insight into design problems influencing production and also made him resolve to return to college with a more balanced schedule and earn his degree.

After graduation, he returned to Texas Instruments to design military products, but a fortuitous response to a Radio Shack ad and successful interview with Leininger led him to his true passion, the consumer. As Chatham recounted,

"The job was exactly what I wanted to do. I like working with consumer products. I like being able to go into a store and see something I designed. I feel lucky because every day I'm doing exactly what I want to be doing." That passion, it seems, would rub off on the CoCo's users.

The magazines of the time reported the rumors that Tandy had been planning to call its successor to the TRS-80 Model I the "TRS-90." These rumors were soon updated to more accurately reflect what would be Tandy's final decision, like in the March 1980 issue of *BYTE*, which reported leanings toward "TRS-80/COLOR."

With the work behind the scenes proceeding apace, it was finally prime time for Tandy to make an official announcement. It was reportedly an appropriately literally "hot day" when many of the short-listed journalists received their call on July 30, 1980, telling them that a press conference would be held the next day to unveil "some new computer products." As vague as that was, within hours, many reporters nevertheless dutifully made their way to Fort Worth to see what magic the number one computer maker had up its sleeve.

By the time the press arrived, Tandy ratcheted up the heat even more. As one invitee, Stan Miastkowski, described in the winter 1980 issue of *onComputing* magazine, "The Radio Shack people played up the suspense as press people gathered outside a locked hotel meeting room. Radio Shack executives came and went talking in hushed tones. When the appointed time came, we were led into a darkened room and the lights were turned up—revealing what we were waiting to see. Radio Shack introduced not one, but three new TRS-80s..." The three new systems were the TRS-80 Model III, the TRS-80 Pocket Computer, and the long-rumored TRS-80 Color Computer.

The TRS-80 Model III was the most pedestrian of the three new systems, since it was simply an improved all-in-one update of the bestselling TRS-80 Model I, designed to conform to the new Federal Communications Commission (FCC) radio frequency interference (RFI) regulations. Although not originally intended as a replacement for the Model I, those same FCC RFI regulations effectively made it one. Prices for the Model III started at $699.99 for a bare bones 4K unit, though the $999 version was far more usable, with its 16K of memory, uppercase and lowercase text, real-time clock, and enhanced version of BASIC.

The TRS-80 Pocket Computer was the most unusual of the three new systems, as it was a rebranded Sharp Electronics-developed calculator-sized, BASIC programmable computer with a small, 24-character alphanumeric LCD screen (Figure 3.2). The price for the Pocket Computer was $249, and included a carrying case, manual, and batteries. It would be the first in a series of market-leading pocket and portable computers, including most famously, the rebranded Kyocera-developed TRS-80 Model 100 introduced in 1983, which would allow Tandy to lead that particular market niche throughout the 1980s.

It was in September, however, that the system most publications of the time thought would generate the most interest, the Color Computer, began to make its formal appearance in Radio Shack stores all over the country (Figure 3.3). These new color computers were available in 4K and 16K models, both based on Motorola's 6809E processor instead of the Zilog Z80, the chip Tandy had used in

Figure 3.2

Author Isaac Asimov pitches the impressive, but relatively limited, TRS-80 Pocket Computer in a 1982 Radio Shack advertisement.

its earlier TRS-80s. Of course, this meant that the name TRS-80—the only major similarity between the Color Computer and its older siblings—had become a misnomer, something that would change in 1984 when then company president Bernie Appel would remove the TRS-80 label in favor of the Tandy brand on all of their computers.

The 4K model was priced at $399 and was upgradeable through a Radio Shack service center to 16K for $119, and then Extended BASIC (which required the

Figure 3.3

The first appearance of the Color Computer in the TRS-80 Computer Catalog, No. RSC-5, 1981. While this section of the catalog started on page 28 with a typical-for-the-era image of the Color Computer being used for professional work, page 29 demonstrates that the new computer was clearly targeted by Tandy for home use. (Courtesy of www.RadioShackCatalogs.com.)

extra RAM) for $99. The 16K model, which came with Extended BASIC prein-stalled, was $599. Though these prices may seem high now, for the time, they were competitive. Competing systems released in the latter part of the previous year, like the 16K Apple II Plus, 8K Atari 400, and 16K Texas Instruments TI-99/4, retailed for $1195.00, $549.95, and $1,150.00, respectively.* Only the 5K Commodore VIC-20, which debuted in June 1980 at the Consumer Electronic Show (CES), and saw wide retail release in early 1981 at $299.95, represented a significant threat to the CoCo's overall value proposition.

Accessories available at the CoCo's launch included the TRS-80 Color Video Receiver for $399.00, which was really just a 13″ color television styled like the computer; the Quick Printer II ($219.00) and matching cable ($4.95), which electrostatically printed 32 characters per line on a 2¾″ wide roll of aluminum-coated paper; the CTR-80A Cassette Recorder ($59.95), which required batteries

---

* While the Apple II Plus's high price was consistent with the company's desire for high profit margins, the TI-99/4's inflated figure came from being bundled with a modified 13″ Zenith color TV. This bundling was the result of failing to get FCC approval for a planned RF modulator that would have allowed use of any television.

or a separate AC adapter, and was used for loading and storing programs on tape using its included cable, which was also available separately and worked with most other standard cassette recorders; the TRS-80 Telephone Interface II ($199) and matching RS-232 Connecting Cable ($19.95), which was a full 300 baud originate/answer acoustic coupler modem; and analog joysticks ($24.95), which came in a pair, did not self-center, and featured a single action button. Naturally, some of these accessories, like the printer, cassette recorder, and modem were designed for or also worked with the original TRS-80 computers, but it was still nice for new Color Computer owners to have immediate access to key add-ons, a luxury that new owners of other early personal computers often had to do without (Figure 3.4).

Like the VIDEOTEX, the new Color Computer featured a gray-colored plastic case. As users would soon discover, the gray paint could rub off over time, particularly in areas that saw heavy use (like where the user's wrist would rest below the keyboard), which exposed the black plastic underneath. In place of the VIDEOTEX Terminal's DATA indicator light on the case was a RAM badge similar to the one found on the Model III that indicated how many kilobytes it had, for example, 4K RAM.

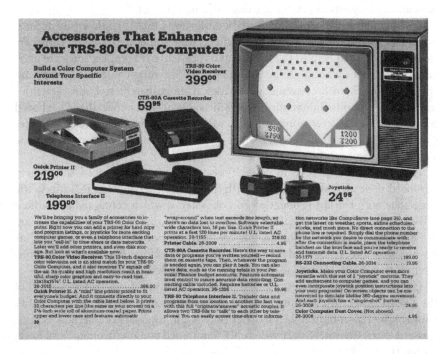

Figure 3.4

Page 30 in the TRS-80 Computer Catalog, No. RSC-5, 1981, which showed many of the accessories available at launch. These accessories were a mix of legacy TRS-80 hardware and newly created peripherals specific to the Color Computer. (Courtesy of www.RadioShackCatalogs.com.)

Since it was not targeted to "serious" users in the way that the Model III was being positioned, the CoCo featured the same plastic chiclet keyboard found on the VIDEOTEX Terminal, with widely spaced keys. This was clearly a step down from the full stroke keyboards found in the earlier TRS-80 computers but was consistent with what was found on many of the other low-cost home computers of the time. In fact, in many contemporary previews and reviews of the CoCo, the keyboard was rarely given much thought, with, for instance, the July 1982 issue of *Popular Computing* magazine casually referring to it as a "button-style keyboard," before moving on to other topics.

Although the CoCo's keyboard may look and feel* odd to modern eyes and hands, respectively, it was common practice at the time to experiment with alternate keyboards that were not necessarily conducive to comfortable touch typing. Not only was there a smaller percentage of people well-versed in touch typing at the time—a skill that was still thought to be somewhat feminine—but the extra costs involved in creating a proper full-stroke keyboard were prohibitive in relation to the costs of the computer's other components. Cost savings had to be realized somewhere and it was much easier to sacrifice the keyboard design, which could save anywhere from a few dollars to $25 or more per unit, than to scale back accessory ports or other electronics deemed critical.

One benefit of the CoCo keyboard's design was that it could easily accommodate keyboard overlay sheets. These sheets were sometimes used on early personal computers to help in the operation of particularly complex software or as an aid to novices who might be intimidated by learning computer or software commands without a convenient reference. Unfortunately, few software titles made use of this possibility, minimizing the value of such a specialized keyboard configuration. However, some utilities, such as Soft Sector Marketing's Master Control (1982), which allowed for single-key entry of complex commands, made a good argument for the option.

Fortunately for those who could type well, reasonably priced third-party keyboard upgrades were available, with Tandy itself attempting to address the deficiency with the last production run of systems, which contained a modified full-stroke keyboard with shallow keycaps similar to that of a modern slim laptop or Bluetooth keyboard. Unfortunately, for many users, this "melted keyboard," as it was not so affectionately dubbed, was still sorely lacking in the typing responsiveness that many of the CoCo's competitor's afforded (Figure 3.5).

Outside of the top-mounted keyboard and the right-side cartridge port, all of the CoCo's other functionality and external ports were to the rear. From right to left were the Power button, Left Joystick port, Right Joystick port, Serial I/O port, Cassette port, Channel Select switch, RF (combined audio/video) out, and Reset button (Figure 3.6).

---

* While it can be argued that superficially the CoCo's keyboard does bear a passing resemblance to today's common chiclet keyboards, the differences in spacing, comfort, and responsiveness are actually quite dramatic. Both variations of this keyboard type are clearly products of their respective times.

Figure 3.5

A "melted keyboard" variation of the original Color Computer. These keyboards were typically found on late model, 64K Color Computers in the white cases, as well as early production runs of the Color Computer 2.

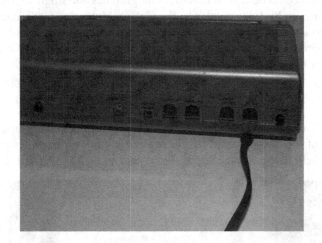

Figure 3.6

The CoCo's rear ports.

Straight from its appearance in Project Green Thumb and concurrent usage in the VIDEOTEX Terminal, the CoCo leveraged the modest, but proven Motorola MC6847 VDG chip for its graphics and text. The VDG certainly seemed flexible enough, supporting one alphanumeric mode, two semigraphics modes, and eight full graphics modes, though it did require direct control from the CPU to generate its analog output. For the CoCo's purposes, the VDG could display as many as nine colors: black, green, yellow, blue, red, buff (near white), cyan, magenta, and orange, simultaneously on a standard color television set with a maximum resolution of 256 × 192, though most software was limited to four colors: black, blue, red, and buff, and a resolution of 128 × 192. Text display was similarly limited to only uppercase characters, though this was a fairly common restriction for the time. Inverse, or reverse video, was used to represent lowercase characters and was never a particularly popular substitute for the real thing (Figure 3.7). Alternatives were available, from software that used the more resource-intensive technique of using graphics to simulate upper- and lowercase character sets, like Nelson Software Systems' *Super "Color" Writer II* (1983) word processor, to internal hardware add-ons and modifications, like the MSB Electronics lowercase board.

Sound output consisted of a single channel or "voice," which was on par with some other computers like the Apple II, but subpar in comparison to newer computers like the TI-99/4, which featured three channels, or the Atari 400, which sported four channels. Like the MC6847, the CoCo's 6-bit DAC, or digital-to-analog converter, required direct CPU intervention to feed audio data to the output. This restriction allowed for greater flexibility in how the graphics and sound could be controlled but also further limited overall performance since CPU time had to be partitioned between the two.

Although it could be argued that the color graphics and single-channel sound were subpar even by 1980 standards, the CoCo's Motorola 6809E microprocessor, in contrast, was a powerhouse. While the MC6809E was an 8-bit CPU, it featured several 16-bit operations in its instruction set, which gave it increased performance over other 8-bit microprocessors. In addition, the flexibility of this instruction set gave its more sophisticated programming languages, like the Extended Color BASIC, higher efficiency than was possible on many competing systems.

Making the expensive but powerful microprocessor the main competitive differentiator of the CoCo was an interesting gamble—flashy graphics and sound were easier sells to the average buyer—but it paid off for Tandy in the devotion of

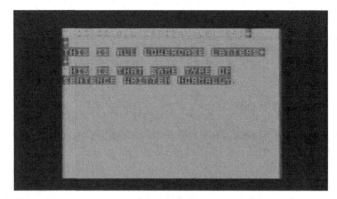

Figure 3.7

A screen capture from Tandy's *Color Scripsit II* (1986), showing how lowercase characters were represented with inverse, or reverse, video.

Figure 3.8

A screen capture of the CoCo's default boot screen, which famously featured a flashing cursor that cycled through all eight of the available colors.

its most technically minded fans through the end of the production run for the series (Figure 3.8).

As Tandy did with Level I and Level II BASIC on its earlier TRS-80 systems, two versions of standard BASIC were offered for the CoCo. The stock 4K CoCo shipped with Microsoft's Color BASIC resided on an internal 8K ROM and was the operating environment that greeted the user when first starting the computer. Microsoft's more powerful Extended Color BASIC, which was well-received for its versatility and power, was offered as a replacement option for use with 16K or higher memory models. Although most were little used or supported, other alternative operating systems were released on disk, including Tandy's version of the powerful OS-9 Level I, which required 64K, but whose potential was not realized until an updated version was released for the more capable CoCo 3.

Within a year of launch, Radio Shack made over a dozen program pak cartridges available. These cartridges included the productivity title *Personal Finance*; the educational and entertaining *Math Bingo*; the utility, *Diagnostic ROM*, which tested system memory and other functions; and, most betraying its consumer-oriented focus, games like *Backgammon, Checker King, Dino Wars, Football, Microchess, Quasar Commander, Skiing*, and *Super Bustout* (Figure 3.9). All of these early titles were designed to run on the base 4K machine, though a few titles, like *Personal Finance*, with its use of cassettes for storage, or *Skiing*, which performed better on 16K machines, pointed the way to upgrades. In addition, most of the games required joysticks.

Several notable programmers moved from working on the TRS-80 Model I and III to working on the CoCo, either in whole or in part. Although the architectures were completely different, the appeal of a new Tandy computer with built-in color graphics and sound was hard to resist. One such early programmer who made a successful switch was Robert Kilgus, who, like most developers of his time, single-handedly programmed everything from art programs to productivity software to games for Tandy.

Figure 3.9

Page 40 in the TRS-80 Computer Catalog, No. RSC-6, 1982, which shows many of the early cartridges that were available for the CoCo. Although cassettes and disks were important to the CoCo's growth as a serious computing option, cartridges would remain a mainstay for the life of the platform, which had, among its advantages in access speed and stability, the important advantage of being the one format every owner with enough RAM was guaranteed to be able to use. (Courtesy of www.RadioShackCatalogs.com.)

Kilgus was always looking for clever ways to make his own programs do more than might be expected. Three of his CoCo launch window titles in particular stood out in this regard: *Dino Wars*, *Skiing*, and *Quasar Commander*. For both *Dino Wars*, which was one of the earliest one-on-one fighting games, and *Skiing*, which not only featured a standard arcade-style mode but also had an advanced mode that took into account the effects of inertia, gravity, and friction, Kilgus set forth the difficult task of making his games speak. This type of speech feature was absent at the time on competing systems without special hardware add-ons, or, at the very least, the extra sound sample storage space disks or cassettes afforded over cartridges. Kilgus's task then was to overcome both the lack of a hardware add-on and the limited ROM storage space (8K or less at the time) a CoCo cartridge afforded.

The solution for Kilgus was to write a clever utility program that allowed him to record and then highly compress a series of square waves with varying pulse widths. As Kilgus remembered, this was no small feat, "I quickly discovered that any word containing an 's' sound required much more space to record than, for example, an 'e' sound. In the word 'set,' all but a small part of the 'recording' was devoted to the starting 's' sound—with a lot of small count values. This led me to another idea. I needed a 'white noise' generator. Perhaps if I could write a small subroutine that could generate a hissing sound, I could use that to produce the 's' sound followed by the recorded data for the 'et.'" Kilgus continued, "The white noise subroutine worked wonders. I started with a random-number generator driven by a Fibonacci series. I used the next random number as a count value for each audio polarity change. That worked but produced a rather high-pitched hiss—like air leaking from a tire. I wanted a more natural sound like when the librarian 'shushes' noisy patrons. The solution was to add a loop within a loop. Each time delay consisted of a count from the Fibonacci and a constant multiplier value. The results were amazing!"

This attention to detail allowed *Skiing* to speak, "Get Ready" and "Get Set," and replicate the sound of a starter's gun, ski-pole hits, and a roaring crowd. This also allowed the dinosaurs in *Dino Wars* to both roar on command and let out a yelp when bitten. As Kilgus recalled, the recording process was almost as much fun as the finished games. "There was one more modification I made to my playback subroutine. After discovering the usefulness of a loop-within-loop to control the pitch of white noise, I applied the same technique to the playback of prerecorded sound segments. That meant I could play back 'ready' and by varying a pitch value, it could come out sounding like a child's voice or a husky man's voice or anything in between—regardless of whose voice was initially recorded. That led to a very interesting week or so of strange sounds coming from the open windows of the Kilgus house one summer. I would hand the microphone to [my wife] Peggy and start my recording utility. Peggy would 'roar' into the microphone. I would then spend a lot of time isolating the roar and playing it back with a variety of parameters. Sometimes, the results were quite funny. Peggy's roar might sound like a small dog's yapping bark, or it might sound like a whale's deep bellow. In between there might be something that could be described as the growl of a bear . . . or even a dinosaur's roar." Kilgus added, "And of course, we were also searching for the sound of a dinosaur yelping in pain from a bite. Many of the unexpected results were followed by the two of us laughing our heads off. I always wondered what our neighbors thought was going on in our house that summer."

For *Quasar Commander*, the end result was a bit more subtle but no less impressive. Much like Atari's high-profile classic *Star Raiders* (1979) for the Atari 400, *Quasar Commander* was one of the first personal computer programs to render an effective 3D world, which required incredibly complex calculations. The player was presented with a first-person view of space from their ship, where speed and careful maneuvering were critical to laser-blasting enemy fighters and landing on a space station for refueling. Objects would appear to scale larger as your ship approached them and would maintain their true location and flight path wherever they were located in space, even if out of the immediate field of view. For its final trick, like *Skiing*, *Quasar Commander* recognized when the user had more than 4K of RAM and was able to use the extra memory to buffer the display, creating an even smoother performance.

Through these early games, CoCo users got a tantalizing glimpse into the seemingly unlimited potential for innovation that their chosen platform could provide. It was clearly going to be a fun ride.

In addition to the cartridges, the CoCo supported an increasing array of peripherals and add-ons, including floppy disk drives and various modems (Figure 3.10). One critical first-party add-on, which was first hinted at in the

Figure 3.10

A TRS-80 Color Computer advertisement from the January 1982 edition of *Personal Computing* magazine describing the value found in adding one or more disk drives.

August 1983 issue of *THE RAINBOW*, was Tandy's Multi-Pak Interface, which allowed users to easily switch between four different program paks. More important, the Multi-Pak allowed CoCo owners to use more than one cartridge simultaneously, such as the floppy disk controller with RS-DOS (Disk Extended Color BASIC), and a speech and sound pak, which greatly enhanced the system's modest sound capabilities when support was programmed for it.

Other accessories to join this ever growing list included high-resolution joystick and mouse adapters; the X-Pad, which was a stylus-driven graphics tablet; the Orchestra-90 CC, which was an 8-bit stereo audio player that plugged into the CoCo's expansion port; and an innovative Electronic Book. More of a wired binder with touch points than an actual book, users would purchase add-on software that came with preprinted pages to place inside of it. When a particular point on a page was pressed, it would then trigger an interaction with the computer.

Compared with contemporary Apple, Atari, and Commodore computers, the CoCo suffered from a dearth of quality third-party commercial software. This was mostly because of Tandy's own restriction that its Radio Shack stores only carry software that it published. If a developer or third-party publisher did not agree to Tandy's sometimes lopsided licensing terms—or Tandy simply turned them away—then there was no chance that their software would appear in one of the thousands of coveted Radio Shack store locations.

The only major alternative then for these third parties was to sell mail order through one or more of the thankfully relatively ubiquitous user groups, newsletters, or computer magazines, both generic and those specific to the CoCo. Although logistically reaching the right audience could prove a challenge, for those third parties who figured out how to make it work, it could also prove quite profitable. These groups and publications played a vital role in building a loyal and dedicated community of CoCo enthusiasts and were instrumental in maintaining a critical market for independent developers.

Because of its versatile design, the CoCo was always attractive to hobby programmers, and, as a result, a great many titles were released for the system by "basement coders," who often sold their wares in plastic zip bags in conjunction with limited but strategically placed advertising. Typical of other computers of the time, the CoCo's software library was also greatly enhanced by the thousands of titles written in its versions of BASIC, including the countless program listings found in various magazines that were just begging to be retyped into the CoCo and learned from (Figure 3.11).

Officially licensed ports of arcade games were few and far between, though knockoffs were popular and plentiful. Tom Mix released two such knockoffs by Chris Latham, *Donkey King* (1982), based on Nintendo's *Donkey Kong* (1981), and *Sailor Man* (1984), based on Nintendo's *Popeye* (1982). Although even the best of these ports and knockoffs tended to look primitive compared with their equivalents on competing platforms, it was ultimately the play that counted, and with that, many of these titles would become beloved by CoCo enthusiasts.

Since sound support was limited to a single 6-bit channel, it made some of the more common features found on other platforms, like background music,

Figure 3.11

Typical of the era, Color Computer software came in every imaginable type of fit, finish, and size.

extremely difficult to code for. Furthermore, the CoCo's slightly awkward analog joysticks, which lacked self-centering, were inappropriate for many types of arcade games. Nevertheless, a few quality, competitive conversions made it through, including from Datasoft—among the platform's biggest official third-party supporters at the time—licensing the oft-ported horizontal arcade shooting game from Konami, *Pooyan*, which Tandy published in 1984.* Other notable third-party partners included Spinnaker and The Learning Company, who brought several of their key educational titles to the platform, including, respectively, *Kidwriter* (1984) and *Robot Odyssey I* (1984).

A few of the early major third-party publishers even handled their own distribution, including Infocom, with classic text adventures like *Zork* (1983) and *Planetfall* (1983); and Avalon Hill, with strategy titles like *Panzers East!* (1984) and action titles like *Breakthru* (1983), which was also listed as being compatible with the TDP S-100. Although there were CoCo computer resellers other than Radio Shack, the Tandy Data Products TDP-100 computer, sometimes referred to as the TDP S-100 (no relation to the bus standard), was a restyled CoCo that was developed for release through alternative distribution channels. It was a

---

* Datasoft hired Steve Bjork to help create their Color Computer gaming department. Bjork's hiring was no accident, as he had previous experience with the CoCo's graphics capabilities from creating color information displays for cable systems using the earlier Electric Crayon product from Percom. Percom's add-on allowed the black and white TRS-80 computers to output a color video display thanks to the same 6847 VDG graphics chip later found in the CoCo.

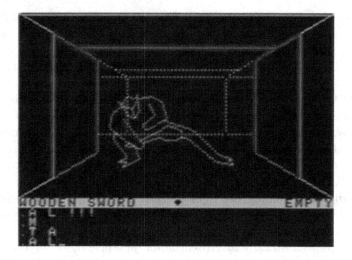

Figure 3.12

A screen capture of a tense monster encounter from *Dungeons of Daggorath*.

short-lived experiment and a relatively small number of TDP S-100 systems were produced, making finding one today a particularly sought-after collectible.

Tandy also invested in cassette-based text adventure games like *Madness and the Minotaur* (1981), *Bedlam* (1982), and *Raaka-Tu* (1982). Text and graphics adventures like *Black Sanctum* and *Calixto Island* were third-party favorites released by the software company Mark Data Products in late 1983. But one of the most innovative and popular adventure titles for the system—and still beloved today—was a graphical dungeon crawler called *Dungeons of Daggorath* (Figure 3.12). This real-time, first-person perspective 3D game was released on cartridge in 1982 by Dyna Micro and published through Tandy. *Dungeons of Daggorath*'s use of sound effects to indicate the proximity of monsters and the avatar's status (via a realistic heartbeat sound) were ahead of their time and greatly added to the player's tension. The player could also string together long lists of commands simultaneously, allowing for rapid, yet sophisticated combat strategies.

Other notable titles included the menu-driven adventure game *Plateau of the Past* (Zytec, 1986); the exploration platform game *Downland* (Spectral Associates, 1983), as well as its more complex sequel, *Cave Walker* (1986); and the definitive adventure game *The Interbank Incident* (Spectral Associates, 1985). *The Interbank Incident* was a true epic for the time, making use of the advanced OS-9 operating system, and spanning three floppy disks, with the rare option to run it all from a hard drive. Its icon-based interface could be controlled via joystick, or, preferably, mouse, and it even made use of the Speech/Sound Pak, if one were present.

Despite all the fun, games, and productivity to be had, for most consumers in the early 1980s, the idea of owning a personal computer was still a difficult concept to grasp, so manufacturers would often go out of their way to try and demonstrate the many practical benefits of ownership. One such tactic erupted into a sort of spokesperson arms race, which was born from an attempt to put a friendlier, more approachable mainstream spin on these sometimes intimidating machines. These celebrity spokespersons included Alan Alda (Atari), William Shatner (Commodore), Bill Cosby (Texas Instruments), George Plimpton (Mattel), Sarah Purcell (Tomy), Roger Moore (Spectravideo), and Dom DeLuise (NCR). IBM sidestepped dealing with the demands of a living celebrity by licensing Charlie Chaplin's *The Tramp* likeness to pitch its PC and PCjr line of computers. Tandy, despite the obvious advantage of thousands of retail locations and national advertising, was not immune to the trend, enlisting the services of newly hooked computer user and prolific author, Isaac Asimov, in late 1981[*] to pitch its latest line of computer products, including the Color Computer (Figure 3.13).[†]

In addition to landing the right spokesperson to help distinguish their machines from the competition, marketing and design teams would work overtime to create a unique "personality" for their products to attract and identify with their owners. An early advertisement for the Apple II, for instance, shows a handsome and stylish young man in a turtleneck creating a graph and sipping from a bright orange cup, while his smiling wife looks on with obvious admiration as she chops tomatoes. An early ad for the Commodore PET, by contrast, shows the computer against the stark backdrop of a college classroom complete with chalkboard, with a stern-looking professor standing at a lectern. Perhaps some early adopters made their decisions based simply on which of these images best suited their personality, though it is more likely they also consulted with salespersons and any other warm body they knew with a computer.

While companies battled it out in the mainstream media, a decisive struggle was taking place in thousands of small groups and between pockets of fans all over the country. Regardless of a platform's perceived strengths and weaknesses, many consumers who invested in them felt a keen sense of loyalty to their chosen brand, especially if they were part of a group of like-minded enthusiasts. In much the same way that football fans become enamored (and defensive) of their favorite team, computer and console owners can grow downright belligerent if anyone questions their choice. Often derided as "fanboys," these extremely loyal and devoted system owners would literally preach the virtues of their system (and vices of the rest) to anyone who cared to listen, and plenty who did not.

It was not all just fanaticism, though. If the kids in your neighborhood owned ColecoVisions, for instance, you could borrow or trade games with them, an important consideration given the high cost of acquiring new cartridges. Likewise, computer owners could copy and swap software, sometimes legally (public domain software or their own programs) and illegally (piracy). If you were the isolated Atari 400 owner in a town full of CoCo fans, you would miss

---

[*] Deborah Wise, "World-famous author Isaac Asimov converts to word processing," *InfoWorld*, January 11, 1982.

[†] In 1984, *The Incredible Hulk* TV star Bill Bixby would be called on to promote the Tandy 1000 and 2000 series computers. While Bill Gates was not a celebrity at the time, in a Tandy ad of the same year he famously promoted the first version of Microsoft Windows on a Tandy 2000.

Figure 3.13

Isaac Asimov describing a great deal on the TRS-80 Color Computer in an advertisement from the August 1982 edition of *Personal Computing* magazine.

out on a lot of "free" software—but you might also feel even more pressure to defend and identify with your machine.

As long as there were plenty of first-time computer and console purchasers in the market, the young computer and videogame industries thrived. Once production finally caught up and then wildly surpassed consumer demand, however, the videogame market suffered what became known as "The Great Videogame Crash." The industry had become drunk on its own rising sales success between 1980 and the early part of 1983. Arnie Katz best described the crash in a June 1989 article in *Video Games and Computer Entertainment* magazine: "Companies acted like sales were guaranteed to double every year till the end of time. Publishers stamped out a dizzying array of new cartridges, far more than consumer demand could possibly support. Therefore, retailers dumped the cartridges they couldn't sell at distress prices. The availability of $5 games ruined the market for labels like Activision, Imagic, and Parker Brothers, who wanted to sell titles in the $25 to $40 range."

While fewer systems were sold in 1983 than in the previous year, cartridge sales were up. By mid-1984, however, the industry hit a ceiling, and sales of both consoles and cartridges fell dramatically—followed by a retail backlash against videogames. The mass media turned on the industry, declaring it dead, and investors pulled out wherever they could. Department stores that had once clamored just to keep games on their shelves were now slashing prices and incurring frightful losses just to get rid of them.*

The media consensus at the time was that consumers no longer needed videogame systems because low-cost computers like the Commodore 64 and Color Computer provided the same entertainment value along with all the other benefits of a full-blown computing device. However, news of the videogame console's death was premature. After the supply chain was mostly cleared of the glut of cut-rate videogames and consoles, the Japanese company Nintendo was able to restore consumer confidence. The introduction of the Nintendo Entertainment System at the end of 1985 began the process of salvaging the videogame console from the bargain bin of history, though Nintendo had to use some unusual and downright aggressive initiatives to win over skeptical retailers.

This era was similarly unkind to most computer manufacturers, as what started out as a wide open field dwindled to a select group. In the lead was Commodore, who had achieved near total dominance of the low-end computer market with unheard-of bargain prices and just enough computing power for both quality games and more serious software applications. For just a bit more money than a dedicated game machine, families could take home a full-fledged computer. The tremendous value of the Commodore 64 led to unrivaled success—indeed, the unit continues to hold the Guinness World Record for the most sales of any single, traditional computer system. However, soon after the end of this era would also see the introduction of higher-end, higher-priced systems that would eventually eliminate the demand for relatively underpowered low-end systems, particularly when computers based around Microsoft DOS (and eventually Windows) would rise to dominance both in the office and at home. As you will eventually see, Tandy and the CoCo platform would play their own key roles in the evolution of this era.

Despite the tortured fates of most of its competition leading into the mid-1980s— thanks to a toxic combination of market factors, including The Great Videogame Crash, Commodore's aggressive pricing, and generally poor planning—the

---

* This backlash was mostly isolated to North America. The rest of the world's markets featured different growth trajectories that mostly avoided the highs and lows of The Great Videogame Crash.

CoCo's future remained secure. Even with the CoCo's relatively modest sales, unlike much of that competition, Tandy was able to keep its endeavors in this area profitable thanks to the computer's fiscally prudent bill of materials, being able to easily leverage both the reach and advertising of its ubiquitous Radio Shack stores, and by ostensibly being the sole provider of its software and accessories. In addition to strategic, but still profitable, price drops, regular upgrades to the core platform, including increases in memory to 64K and an improved keyboard, also helped keep the platform relevant through this tumultuous period.

An example of Tandy's continuing focus on cost, the 64K Color Computer was a result of one engineer's way to bring the cost down on the hot-selling product. After Paul Schreiber graduated from Texas A&M in 1979, he spent a short time at Data General, then worked as an intern at Tandy before coming on board full time there in June of 1980. One of his first assignments was to design a stand-alone modem that could work on the CoCo and Model I. "'Cost reduction' was a way to get Tandy's attention," recalls Schreiber. Radio Shack would set a retail price, then a spreadsheet was used to back calculate from the target price the price of parts to meet the cost objective. According to Schreiber, "They could not go over by a penny. Bernie Appel would not approve unless it was at or under budget."

Twice a year, Tandy Electronics would present ideas to be approved before Radio Shack's president, Bernie Appel. "It was an internal trade show of sorts, about one or two days, and very intense. Engineers would think of things to present," recalls Schreiber. In a measure of faith, Schreiber put in for a "Color Computer Cost Reduction" as his idea, not really knowing how he was going to do it. The project was approved.

Schreiber began to study the schematic and noticed that the circuit board in the CoCo was quite large, taking up the entire 18″ × 24″ panel. The more board that could be designed on a single panel, the more of a cost reduction could be realized. The ambitious engineer spent three to four days working to get two boards per panel using a "puppet," a 1:1 cutout on gridded Mylar film, then working atop a light table to see how to make the parts fit. Schreiber ended up with a board that was shaped like a capital L.

Schreiber remembers: "You could only use 17″ × 23″ because you needed 1″ around the outside edges, known as the 'dead area.'" The design left a rectangular space in the middle, so Schreiber divided those in half and used those for the AC transformer. In the end, Schreiber's clever work cut the cost in half for the raw circuit board on the CoCo.

Another cost reduction that Schreiber took on was the trim pot, a manual adjustment for the voltage. This was adjusted at the factory for each assembled Color Computer. The process took time and human labor, and was done because the regulator was only accurate ±7%, while a less tolerant ±3% was actually required. Schreiber asked Motorola to test a chip to ±2%, which eliminated the need for the trim pot. It turned out that the cost for doing that was about the same cost as a trim pot, so it was a wash. But the cost of labor was saved, as well as possible human error from using a volt meter to manually set the trim pot for the appropriate voltage.

The other cost optimizations involved having memory selection jumpers for different memory sizes, sourcing different capacitors, and so on. It was exhaustive work but was necessary to meet the goal.

Schreiber then sent the circuit board out for a new layout. When the board came back, he was focused on making sure he didn't break anything with the updated design and that the new Motorola voltage part was working correctly. Everything seemed OK.

With another board spin complete, he turned his attention to the video output, using a factory test cartridge which produced test bars, speaker beeps, and provided a keyboard test harness. Based on the screen tests, Schreiber believed that the video could be improved, so he began to look at how to do just that while continuing to reduce cost. "The RF modulator was bought from a company called ASTEC," recalls Schreiber, who noticed that noise caused by bad grounding and traces on the circuit board was getting into the modulator. It just so happened that another Japanese company was vying for Tandy's business. The company called on the engineer, telling him that it could build just as good a modulator as ASTEC for less cost. Schreiber was open to the idea, noting that while carrier bleed through (color burst rejection) done by passive filtering inside the modulator was acceptable, it was not considered, so for the rest of the design that was being considered.

Schreiber noted parts in the current modulator that he felt were not needed, so he worked with the Japanese company to create a new design based on his new spec. Upon receiving a sample modulator from the company, he tuned the part, installed it, and then saw the video output. "It was gorgeous! Everyone could not believe how great the modulator was," recalled Schreiber.

"Mark Siegel had a reputation for being extremely picky. I thought the new CoCo revision was just fine and would pass. A couple of weeks go by and I get a message from Mark Siegel that the QC has failed and the new design was rejected." Confused, Schreiber asked Siegel for clarification.

Siegel responded: "It doesn't run our most popular game, *Clowns & Balloons*." It turned out that the game was so popular that there were no cartridges available around the office, so Schreiber visited the internal software department to burn a ROM for the game just to get it to test. When he got the ROM installed and plugged it into his low-cost CoCo, it played it just fine.

Even more confused, Schreiber went to Siegel's office. "It's working fine," the engineer confidently pointed out as the game played on the television. Siegel responded: "Where's the red and blue colors?" Schreiber did not hesitate: "There aren't supposed to be colors; it's in hi-res mono mode."

Siegel walked over to an older production Color Computer and plugged in the cartridge, whereupon the pseudocolors of the game could be readily seen. "See, that color right there," Siegel indicated. It then dawned on Schreiber that his changes to the RF modulator to optimize and clean the video had inadvertently removed an unintended "feature" that had become a staple for games.

"That was a very slow and lonely walk back to my office," says Schreiber, who pulled the data sheet for the 6847. With more study, he realized that the CoCo had "chroma leakage" into the RF modulator. The lack of rejection circuitry was in part to save cost, but a side effect was generating the false color modes. Ironically, by making the video better, Schreiber had actually removed an important feature, one that he was on the losing side of the argument.

According to Schreiber, a meeting was called, which included Barry Thompson, Van Chandler, John Prickett, Dale Chatham, Chris Kline, and the manufacturing folks. It was decided that since the mode had been used in *Clowns & Balloons* and other games, and since other game developers were using it, Schreiber would have to inject noise and leakage back into the circuit. The young engineer consulted with his boss, Chris Kline, who told him it was best to build a band-pass filter (with a resistor, capacitor, and inductor), costing about 11 cents for Radio Shack (which negated his cost reduction for the power supply) and inject, on purpose, the color burst signal at a very low level (millionths of a volt) to cause the false color to come back. The injection was referred to jokingly as "improving the badness" but had an interesting side effect; it actually made the colors even brighter than before. As seemed to be the case with everything CoCo, it was this incredible, somewhat obsessive, attention to detail and innovative thinking that would continue to drive the platform forward.

While an incredible amount of work was proceeding mostly unseen behind the scenes, Tandy still had its ace in the hole, Radio Shack, as its everyday face to the customer. With the dizzying array of computing platforms coming and going on a regular basis, it must have seemed like finding the proverbial pot of gold to the CoCo's influx of new owners to have a relative oasis like a Radio Shack to turn to in an increasingly arid personal computing landscape. Longtime CoCo fans, however, knew that the guiding rainbow to that pot of gold was not solely a Tandy creation but also equal parts the efforts of a great number of outside enthusiasts and visionaries, including those of one Lonnie Falk and his aptly named magazine, *THE RAINBOW*.

4

Somewhere over
the Rainbow

By the time the final issue of *THE RAINBOW* magazine was published in May 1993, it had served 155 consecutive monthly issues to hundreds of thousands of Color Computer enthusiasts for over 11 years. In contrast to most of today's slick computer magazines that place a premium on white space and novices, *THE RAINBOW*'s pages mirrored that of its best contemporaries. It was packed with pages of detailed information that appealed not only to new users but also expert hobbyists and engineers, reflecting the demographic of the era's pioneering computer owners. During its lifetime, *THE RAINBOW* taught readers of all ages how to use their CoCos, helped to foster the creation of numerous software and hardware businesses, and employed dozens of talented individuals. Best of all, like many good success stories, it started out modestly, in an upstairs bedroom on the outskirts of Louisville, Kentucky, in 1981.

Before Lawrence "Lonnie" C. Falk made Color Computer history with the success of *THE RAINBOW*, he was part of history of a different sort. Born in Mountain Brook, Alabama, Falk attended the University of Alabama, where he witnessed the infamous "Stand in the Schoolhouse Door" made by Governor George Wallace at the onset of the university's historic antisegregation integration effort in 1963. After receiving a degree in communications, Falk set out on his career, becoming a journalist for United Press International (UPI). While at

UPI, he reported on some amazing stories, including NASA's first moon landing and the historic Category 5 winds and raging destruction of Hurricane Camille (Figure 4.1).

By 1980, Falk had built upon his impressive journalism career and settled into managing the public relations department at the University of Louisville, where he worked with news bureau editor Jim Reed. Although they did not realize it then, both men would have a profound impact on the CoCo community in the years to come.

Falk's introduction to the Color Computer could not have come at a more opportune time. In that early period when interest in the CoCo was first growing, there was very little in the way of available information about this new marvel of a machine. Hungry for information, new CoCo users were looking for ways to get the most out of their computers, while also connecting with other owners. Like these users, Falk was also enamored with the CoCo. As he began to read the manual and write BASIC programs, Falk became interested in what else the CoCo could do and began to think about how information on the computer could be disseminated and shared with other like-minded individuals. Using his newspaper roots as an inspiration, Falk began the process of making a newsletter, turning out several issues on his own before asking his colleague, Jim Reed, for assistance in moving to a magazine format. Reed, an accomplished writer who

Figure 4.1

Lonnie Falk as a UPI reporter at the scene of the Lunar Receiving Lab where Neil Armstrong, Buzz Aldrin, and Michael Collins were returning from their historic Apollo 11 mission.

spent a considerable amount of his time creating and editing press releases for the university, agreed to help.

The first formal edition of *THE RAINBOW*, as *"the RAINBOW,"* was created and christened in the July 1981 issue, Volume I, Number 1 (Figure 4.2). Printed on a Radio Shack Line Printer VII dot matrix printer, the issue was composed of just four pages of information printed on both sides of two sheets of paper. It was a modest effort with a scruffy, colorless appearance, but it would be the start of something big under Falk's Falsoft company moniker, an amalgamation of "Falk" and "software."

Vol. I
No. 1

the RAINBOW

5803 Timber Ridge Drive • Prospect, KY 40059

First of all, welcome to the RAINBOW!

We'd like to open things up for volume one, number one with a little background on why the RAINBOW exists and what we hope to accomplish with this newsletter. If you are anything like us, we think you'll see that this publication is well worth the small amount we'll be asking you to invest in it.

Most of us are among the first to be the proud owners of a TRS-80 COLOR Computer. And, if you are like we were, you were attracted to TRS-80 in the first place by all those great programs available for the Models I, II and III.

But, where did that leave us? Except for some programs in the manuals -- and the e-x-p-e-n-s-i-v-e ROM Packs offered by the Shack -- there just isn't a great deal out there right now. Oh, it is coming. But the wait seems long and there are a lot of things the COLOR Computer can do that its big brothers can't.

And, frankly, what software is available might be good or might be bad. I'm sure you, like I have, hear all sorts of stories about software that doesn't work, won't load, and so forth. One of the things we expect to be is YOUR representative to the software distributors. We'll be doing reviews and we'll give you the straight scoop. We feel confident that those dealers who have a good product -- and who want to reach an interested and receptive market -- will be happy to let us review their software. And, be sure, you'll get a fair and honest appraisal from the RAINBOW.

We'll also pass along programs on these pages. Those we come across ourselves, and those you may want to contribute. Don't be bashful. And don't think your program has to be a "monster" to get reproduced here. (In fact, as you'll see shortly, the program used to write the RAINBOW takes up just three lines.) So, send in your favorites and we'll share them with the world.

One most important thing. The RAINBOW is devoted to the COLOR Computer and the COLOR Computer only! While future issues will share some of the "secrets" of program conversion, you won't find anything here about any other computer. No Mod III stuff, no Apple stuff, no Atari stuff. The RAINBOW is going to be the one -- and, to our knowledge ONLY -- publication that's just for COLOR Computer owners. In other words, we think you'll find everything you you read in the RAINBOW will have a direct application to YOUR computer.

Ah, the commercial. The RAINBOW costs money to print and mail. In order to pay some of those costs, we're charging $12 a year for a subscription. We hope you'll find that reasonable and that you'll send a check by return mail. Frankly, that's pretty cheap for all the information you'll get. I hope you'll agree. Thanks a lot for listening. We look forward to hearing from you.

Figure 4.2

The cover of the very first issue of *"the RAINBOW,"* July 1981.

Each month, the readership as well as the size of the newsletter grew. With the publication's unexpected growth, Falk found himself spending more time at the local drug store making copies. It soon became apparent that this increase in subscriptions would require scaling the production method substantially.

Realizing the potential his fledgling magazine had, Falk began investing in both printing equipment and people to put out a more professional, polished product. One of his first hires was Courtney Noe, to act as associate editor of the magazine. The staff began to grow to the point where Falk himself quit the University of Louisville altogether to focus completely on *THE RAINBOW*. At the end of 1982, Reed would also leave the University of Louisville to become a full-time employee and *THE RAINBOW*'s managing editor. In his new role, Reed would select and edit the increasing number of articles that came into the magazine to make sure that they were properly structured and grammatically correct as well as oversee the publication's layout, artwork, and printing.

It was there, in Falk's home basement, that Reed recalls the decision was made to standardize the moniker "CoCo," a shorthand name for the Color Computer that was commonly attributed to Dave Lagerquist, who ran cassette-tape-based *Chromasette Magazine*. The inclination in some parts of the fledgling Color Computer community was to refer to the system as the "80C," whereas others called it "CoCo." Both Falk and Reed disliked the former and decided that *THE RAINBOW* would exclusively use the term "CoCo" from that point forward. This would cement common usage of "CoCo" within the community ever after. "The 'CoCo community' had a nice feel to it," as Reed recalled the phrase.

It also became clear that the magazine's base of operation, the 1200 square foot basement in Falk's home, was not going to accommodate the increase in staff and circulation. In addition to the increasingly cramped space, the late night work sessions in Falk's home to put the magazine together were getting to be more than a minor inconvenience for Falk's wife, Willo, then a schoolteacher.

The start of 1983 marked a huge step for Falsoft. Falk took his growing business out of his house and into a 2100 square foot storefront in Prospect Point Shopping Center, a strip mall on the outskirts of Prospect, Kentucky—and only a short hop from his home. The space, formerly occupied by a beauty parlor, was housed between the town's post office and a pharmacy. Falk and company (Figure 4.3) continued to grow the magazine and pump out monthly copies of *THE RAINBOW* to thousands of eager CoCo fans from around the country.

Improvements to the magazine came at a swift pace. By the July 1982 issue, the Line Printer VII was retired in favor of a professional typesetting system and an appropriately full-color front cover. The January 1983 issue brought even more changes, such as a switch to glossy pages to join the case/perfect binding from the month before (no more staples!), and, for the first time, crossing the 200-page mark thanks in part to an uptick in the number of advertisers.

The January 1983 issue also featured a cover with the debut artwork of illustrator Fred Crawford, a name that would become synonymous with many of *THE RAINBOW*'s covers over its long life. A graduate student at Falsoft's

Figure 4.3

Falsoft employees Jim Reed, Courtney Noe, and Dan Downard. (Courtesy of Dan Downard.)

favorite recruiting center, the University of Louisville, Crawford also worked in the university's athletic department and drew portraits on the side.

Influenced by famed illustrators Norman Rockwell, N. C. Wyeth, and Maxfield Parrish, Crawford considered himself a "mercenary artist" who worked for hire. Instead of a pistol or carbine, Crawford's weapons were pencil and canvas. His artwork for the university's football team calendar was good enough to catch the eye of Falk, who happened to be around the athletic director's office one day. Upon learning that Crawford did the calendar, Falk began to chat him up.

"I've got this little newsletter that I'm thinking about taking it further," Falk explained. Crawford listened to Lonnie's pitch about THE RAINBOW magazine and the need for some cover art. Intrigued, Crawford agreed to sketch a cover on an "advance against draw" basis, whereby Falk would withhold full payment unless circulation of the magazine reached certain thresholds. It was a gamble for Crawford, but it was worth the risk. As Crawford recalls, more often than not, the numbers hit their mark, and he received full pay for his artwork.

Crawford's illustrations would grace the covers of THE RAINBOW for years thanks to this positive working relationship. To start the creative process, Crawford would visit Falsoft's office and meet with Falk and Reed to receive an idea for a particular issue. A scene that would complement the theme of the issue would vary in its appearance, be it telecommunications, games, or productivity (Figure 4.4). Crawford would then take the idea and sketch it out. Sometimes a live subject would sit and let Crawford patiently draw him or her; other times, the subject was more abstract. From Falsoft employees to the neighbor's kid, Crawford drew from a variety of sources to create his famed magazine cover

The cover shown reads:

February 1986 — Canada $4.95 — U.S. $3.95

Just for your TANDY COLOR COMPUTER

The RAINBOW

THE COLOR COMPUTER MONTHLY MAGAZINE

A Toast to the Utility Program!

Recover Disk Directories

Auto-Execute Tape Programs

Speed Up BASIC Searches

How to Use CoCo's 'Serial Port'

Also
Play Bubble Wars
Join the Commandos, and
Do the Robot Workout

Plus Learn double-speed tape loading, send electronic valentines, create OS-9 system disks, transfer picture files from one format to another, and clip out our Delphi command card.

Figure 4.4

Fred Crawford's whimsical illustrated cover from the February 1986 issue of *THE RAINBOW* showing Falsoft's Tamara Dunn and her startled reaction as a 5.25″ floppy disk flies out of a toaster.

art. Using nothing more than Prismacolor pencils on textured paper, he even drew himself as a subject for the April 1987 magazine cover. It was, according to Crawford, "The cheapest model I ever used."

As Crawford learned, drawing for a technically oriented magazine with a willful readership like *THE RAINBOW*'s had its interesting moments. For instance, one particularly eagle-eyed Australian reader scoured one of Crawford's covers, which featured a drawing of a motherboard, to discover that an incorrect electronic "part" was used. The thoughtful reader contacted managing editor Reed to make him aware of the infraction. Somewhat tongue-in-cheek, Reed told Crawford that subsequent work would have to be vetted for technical correctness.

For a good part of its life, THE RAINBOW magazine's cover was graced with Crawford's Rockwellesque drawings and illustrations, but as time went on schedules became tighter and more compressed. Payment milestones were also hitting with decreasing regularity. Feeling that the quality of the work was starting to suffer as a result, Crawford put down his Falsoft pencil and moved on to other ventures. Nevertheless, Crawford's work remains a critical piece of THE RAINBOW's legacy, as his time with the magazine coincided with its most successful years.

This "golden era" for Falsoft and THE RAINBOW was evident by the July 1983 issue, when circulation increased to over 50,000 paid subscribers and page count to over 300. The timing of this growth was fortuitous, because the popularity of other top newsstand competition, like Hot CoCo, which ran until February 1986, and Color Computer Magazine, whose last issue was October 1984, also peaked around this time as well.

A few months earlier, Falsoft's very first RAINBOWfest was held in Chicago, Illinois. By most reliable estimates, 11,000 to 12,000 people attended this special event, where vendors lined up to sell CoCo wares, and seminars on all sorts of topics were presented. Riding high as they were, Falsoft even embarked on another magazine venture at the time: PCM (Portable Computing Magazine). The only downside to all this success for THE RAINBOW fans was that with the magazine's growth came steady increases to the newsstand price, reaching, for instance, $3.95 by the September 1983 issue, which represented a $1.00 increase over the previous month.

The September 1983 issue also heralded the introduction of the new 64K Color Computer and the Color Computer 2, with Fred Crawford's signature illustration style showing the all-white computer literally leaping out of the cover. Technical Editor Dan Downard penned an enthusiastic review of the new computers, along with Radio Shack's announcement of OS-9 as the new operating system, which would borrow many useful features from the powerful but commercially challenged Unix, including multitasking and hierarchical file directories.

Although there were not many downsides to Falsoft's continued growth in lockstep with the CoCo market, logistics were beginning to prove a challenge. It had gotten to the point where both sides of the Prospect post office were taken up by THE RAINBOW and its sister magazine, PCM. An additional 2850 square feet in an adjacent building was also being used to house employees and production equipment. At the height of the business, the headcount at Falsoft had grown to 70 people, many of whom had no magazine experience. Notes Reed, "Most people were hired for talent and potential, not expertise." In addition, Falsoft continued to rely on the considerable CoCo knowledge from contributors scattered across its readership.

In November 1985, the move to a new two-story (three floors counting the basement) "Falsoft Building" began (Figure 4.5). Though Falk did not own it,

Figure 4.5

Falsoft employees smile for a group photo in front of the Falsoft Building in the summer of 1986.

he successfully negotiated language in the lease contract that mandated that the building be named after his company. At first, Falsoft leased the entire second floor, but eventually Falk's company subsumed part of the first floor and basement, where back issue storage, shipping, and tape/disk duplication took place. The basement also held the process camera for making "camera-ready art" for sending to the printing company.

A considerable portion of the second floor was dedicated to the elegant and softly lit setting of Falk's office, or penthouse, as one might more accurately describe it. An eye-catching fish tank was against one wall, with Falk also enjoying a private bathroom and hot tub, as well as an "escape route" where he could leave the building undetected. It was in this office that Falk penned *THE RAINBOW*'s monthly editorial column whose title, "PRINT #-2," was a coy reference to the CoCo's BASIC command to print text to a printer. In "PRINT #-2," Falk would opine on a variety of subjects, not just the Color Computer. It was also from this same suite that Falk would administer the general direction of the magazine as well as all of the other periodicals that Falsoft would come to publish, including *Soft Sector*, *Scorecard* (for University of Louisville sports fans), the short-lived *VCR Magazine*, *New Pilot Magazine*, and the aforementioned *PCM*.

Managing Editor Jim Reed kept busy handling the day-to-day editorial operations of the company. Although much of his attention was spent on serious matters, he would also take time to read many of the letters submitted by readers. Correspondence ranged from requests for help on a certain program to ideas for new articles to employment queries. Reed recalls receiving a nine-page handwritten letter in the mail—on legal paper, no less—from a young man enumerating the many reasons why Falsoft should hire him. On the last page, he signed his name and followed up with a postscript apology for handwriting the letter because he had to "save the printer ribbon for more important stuff."

Reed was not averse to taking calls from subscribers either, occasionally helping out readers directly over the phone. In one instance, a woman called to complain that her *RAINBOW on Tape* cassette was not working. Reed politely asked her if the tape recorder was plugged into the computer and into the wall; as it turns out, it was not. Another reader phoned Reed to say she was upset that her CoCo was displaying, "HELLO MARGARET," after typing in a BASIC program listing she had found in *THE RAINBOW.* "My name is Ruth!" the frustrated woman told Jim. Realizing that she had typed "MARGARET" verbatim as the example showed, instead of replacing her own name as instructed, Reed patiently suggested: "Type in R-U-T-H, and it will say 'HELLO RUTH.'"

Reed clearly enjoyed keeping in touch with readership of all experience levels. "As *THE RAINBOW* evolved, I held steadfast in the conviction that we would never, ever abandon the beginner—for each month more and more rank newbies would become readers and subscribers. I knew how excited my own first CoCo was, and I wanted others to discover this new machine without feeling overwhelmed by its potential. We must remember that owning a home computer in those early days was about as offbeat as being, say, a ham radio operator."

Several outspoken voices among *THE RAINBOW*'s advertisers and devoted hobbyists argued that the magazine's articles should favor more advanced content as the core readership became more knowledgeable, but Reed insisted on having the preponderance of content remain directed to novices. "New fans were jumping on the wagon every day! While some longtime readers were becoming more sophisticated computer users, most were less immersed. In other words, the vast majority of the CoCo community were new kids on the block."

Other publications did, indeed, spring up with more technically oriented content that attracted some of the more ardent CoCo users, but Reed resisted. "I was willing to concede that end of the market if we had to in order to remain true to the new CoCo owners. Those other publications did cover ground we did not, but we kept true to the casual, fascinated new user and always made sure there was plenty for the less technically inclined."

"As I once discussed with Lonnie," continued Reed, "we could not afford to let the advertisers drive the content; those folks were much more deeply immersed in computing than our typical readers." Even some of the contributing writers, at times, he felt were writing to impress each other, "and I felt a responsibility to not let *THE RAINBOW* evolve into some esoteric technical journal, catering only to the small but vocal advanced users."

"Lonnie and I both took some heat for that commitment to beginners," continued Falk's former editorial chief, "but I strongly believe that commitment to always keep the beginner foremost in our editorial content is what made *THE RAINBOW* enjoy a long run of success. I suspect many would-be readers picked up those other publications and then laid them right back down when they found little to attract a new user. From a business perspective, we reasoned that while those who became CoCo fanatics and craved more 'techie' material might well subscribe to other publications, our belief was they would not abandon their subscriptions to *THE RAINBOW* to do so."

This continual pressure for more in-depth articles was not entirely ignored, however. Reed decided that *THE RAINBOW* was big enough at 300-plus pages to accommodate both the newbie and the seasoned user by offering more advanced technical content, but only by keeping it separate from the main editorial mix. "That's why we came up with the Rainbowtech section. We wanted it to be clear this was only for more advanced users."

As company editorial director, Reed was well aware that one of the Falsoft publications he oversaw, *VCR Magazine*, had an X-rated section that was sealed during the printing process. "That section could be unsealed, but it could also be easily ripped out and tossed if a reader didn't care for that explicit material. Clearly, we didn't need to go to those lengths with Rainbowtech, but I did want to keep the tech side sequestered toward the back of the magazine. It wasn't in a plain brown wrapper, but Rainbowtech was our hardcore section and not for everybody."

Although advertising is the lifeblood of any magazine, good circulation is the key to attracting those advertisers. In turn, to attract and keep those readers, you need solid, interesting content. A vital resource for this content came from the readers themselves. "There are millions of people who want to be published," notes Reed, "and we got our share of lightweight fluff, such as '10 reasons why I will never own a computer.' In my standard rejection letter and in the annual *Writer's Digest* guidebook for writers, I stressed, 'If you don't already read our magazine, don't bother submitting articles. We are a publication supporting one specific model of one specific brand of computers, and we are not interested in general interest or humor articles.' Thankfully, we received a wealth of material from our ever-expanding readership. [However,] fewer than one in a dozen were actually used in our various magazines."

Submissions from readers helped fill out the pages and gave the periodical much of the content that made it what it was. One of the monthly tasks at Falsoft was to sift through the mounds of mail and rank the contributions from readers to see which would make the cut for an upcoming issue of *THE RAINBOW*. Most articles were submitted on speculation, that is, freelance. After each submission was graded, the A's and B's would be looked at for potential inclusion in the magazine's future editorial mix. First, a passing submission was examined for completeness. This "technical review" usually fell to Falsoft employee and technical assistant Cray Augsburg, whose cubicle-sized office was actually once a closet adjacent to Reed's corner office.

Articles could contain photos or software listings in BASIC or assembly language, or sometimes simple one-liners. Depending on the quality and content, an article could fetch anywhere from $100 to $300 or more. At one point, Reed remembers a young fellow who sent in a letter asking for *THE RAINBOW* to "bid" on four PEEKs and POKEs that he discovered. Reed, amused that the young man would request him to bid on a submission, sight unseen, wrote back with the offer: $0, $0, $0, $0.

Copyediting for grammar, punctuation, and consistent style as opposed to technical editing was also an important task, as *THE RAINBOW*'s senior editor Tamara Dunn recalls. "The BASIC listings had to be proofread. On one occasion, there was foul language." In that particular incident, the obscene word was missed in the editing process and inadvertently got published. After the error came to light, the job of editing became much more tedious in order to prevent another oversight. A typical article might be initialed "OK" by more than a dozen people before being given the final approval at a formal "paste-up board session" before going to the printing company.

In addition to ad hoc submissions, *THE RAINBOW* had a pool of regular contributors who were counted upon to write a series of articles. Tony Distefano, of the CoCo hardware company DISTO, had his popular "Turn of the Screw" monthly columns, while Fred Scerbo would focus on educational topics. There was also Marty Goodman's "CoCo Consultations" Q&A column where readers would submit questions to him on various technical topics. Dale Puckett's "KISSable OS-9" column was a hot spot for the more technically oriented OS-9 CoCo crowd.

Besides technical information, *THE RAINBOW* also had a flair for the creative. Falsoft employee Jerry McKiernan graced the pages of the magazine with a series of cartoon strips based on the lovable CoCo Cat, a mascot of sorts for the Color Computer community (Figure 4.6). The CoCo Cat character also spawned a life-sized costume worn by Angela Kapfhammer and other Falsoft employees at various RAINBOWfests. Logan Ward of Memphis, Tennessee, contributed his graphic skills to the magazine with a monthly cartoon titled *Maxwell Mouse*. Although not a Falsoft employee, Ward's work was featured quite often in *THE RAINBOW*, and he received sage advice from Falk. "Lonnie encouraged me to copyright my work," recalled Ward, whose strip was actually drawn using Colorware's *CoCo Max* (1985) graphics program. Ward would also oversee "The CoCo Gallery" graphics competition at RAINBOWfests, where participants could submit their own CoCo-generated art just like they did for the regular magazine department (Figure 4.7).

Reed communicated with the contributors by phone on an ongoing basis and developed trust in their work. While some articles ended up with revisions due to editing, Reed remembers one writer whose work would usually require few changes. "I generally took what Bill Barden would write as gospel." It wasn't surprising given William Barden's impressive body of work. As a writer, Barden authored a number of computer books for Radio Shack, including the famous 1983 publication, *TRS-80 Color Computer Assembly Language Programming*. Barden's articles in *THE RAINBOW* were some of the most highly anticipated and acclaimed. "Barden's Buffer" was the name of the column where Barden himself delved into both software and hardware topics. In many respects, Barden was considered the CoCo community's very own professor.

CoCo Cat

Figure 4.6

CoCo Cat proved to be an endearing character for *THE RAINBOW* magazine and the CoCo community in general. This strip appeared in the December 1986 issue.

A few among the in-house staff also made regular monthly contributions. One of those was technical guru and Color Computer enthusiast Ed Ellers, whose "Earth to Ed" column was titled with tongue firmly in cheek by Reed himself. As Reed recalls, Eller's genius was trumped only by his tendency to wander off on tangents, often forcing Reed to call out, "Earth to Ed," in order to get him back to focusing on the task at hand. "Ed was truly a brilliant fellow," recalls Reed, "but like many visionaries, he often was on a distant plane all to himself. Those of us who worked at Falsoft all have Ed stories, as he was quite a character. Most certainly, he marched to a different drummer. I have dozen upon dozen of fond stories of our days at *THE RAINBOW* and Ed is central to most all of them."

The relationship between Tandy and Falsoft was comfortable, if not cozy. Tandy's marketing executive Ed Juge paid several visits to the home of *THE RAINBOW*, as did Mark Siegel and Barry Thompson. Collaboration between the companies even went beyond Radio Shack's attendance at RAINBOWfests. At

Figure 4.7

"The CoCo Gallery" department was a popular recurring feature in *THE RAINBOW* magazine. Art enthusiasts of all ages would often send in works of surprising quality, as evidenced by the submissions shown here in the July 1987 issue.

Reed's behest, Tandy agreed to do something that was quite a departure for it at the time: inserting a complimentary issue of *THE RAINBOW* in Color Computer boxes destined for stores. The exposure worked well for Falsoft, as it made new CoCo owners immediately aware of the monthly magazine.

Surprisingly, Lonnie Falk seldom saw eye-to-eye with Jim Reed, an assessment that many at Falsoft at the time knew all too well. Despite their differences, Falk and Reed shared a philosophy: If you both agreed on everything, then one of you is not needed. "Make no mistake, the magazine was his baby," notes Reed, "and I suppose I was the 'baby's' chief nanny."

"We have our very own J. R. [Ewing]," Lonnie's wife was known to comment, alluding to the character on the then-popular television show *Dallas*. This give and take between the two over editorial matters kept fresh ideas flowing and contributed greatly to the success that *THE RAINBOW* enjoyed. Reed also carried a philosophy all his own: It was better to make any decision than to not make a decision. When a new idea came up, it would either be tried or discarded. "There were no lengthy committee meetings or talking things to death. We had both come from a university environment where action and change could be painfully slow."

One general staff brainstorming session did prove to be a little too much, as Reed laughingly recalls. It was customary for the editorial, advertising, and business staffs at Falsoft to meet in the conference room once per week (Figure 4.9). One time, in a departure from the norm, Falk wanted a new ideas session. Falk

Figure 4.8

T&D subscription's booth at a RAINBOWfest. With the right product mix, vendors could make lots of money over the course of the three-day event. (Courtesy of Tom Dykema.)

Figure 4.9

Lonnie Falk stands at the head of the conference table in Falsoft's meeting room.

opened the meeting: "No holds barred. There's no idea too crazy. We're going to have an open session with new ideas and other profit centers. No matter how outrageous the proposal." No one said anything at first. "Who has a suggestion? The floor is open," Falk encouraged. After a few moments, one woman piped in, "Maybe we should start a bakery on the first floor." Falk sank back in his chair, covered his eyes and blurted, "That is the stupidest idea I've ever heard." Stone silence fell on the room for a moment, after which Reed broke the awkward silence with a chuckle, and announced, "Well, I guess that's the end of this meeting."

Frankness not withstanding, Falk is remembered fondly by former Falsoft staff members. Senior editor Tamara Dunn recalls, "Lonnie was an impetuous genius. We had so much fun. Everyone sensed that we were onto something big. It was an exciting time in the computer industry."

* * *

Spectral Associates of Tacoma, Washington, was one of the many advertisers in *THE RAINBOW* that brought a plethora of games and software to the Color Computer market. The software distribution company was run out of a house in the South End of Tacoma and owned by Thomas Rosenbaum, who also published *Commander Magazine* for Commodore 64 enthusiasts. Rosenbaum was also the author of the definitive "Unravelled" series of books that contained annotated disassemblies of the Color Computer's Standard, Extended, and Disk BASIC ROMs, and, later, the CoCo 3's Super Extended BASIC. Although most of the software that Spectral Associates sold was written by outside contributors, there were some in-house developers, among them John Gabbard and David Figge.

After buying his own Color Computer and learning BASIC, Figge wrote a game, *Maze Escape*, which he pitched to Spectral Associates for possible sale. They liked what they saw and agreed to not only sell the game but also teach David 6809 assembly language if he would agree to write games for them. Figge went on to write a *Frogger* clone called *Froggie* in 1983. Impressed with the results, Rosenbaum offered Figge a full-time job writing games in March 1984. By this time, Spectral Associates had built a working relationship with Radio Shack. Eager to get its work distributed through Radio Shack's many stores, Spectral Associates tasked Figge with writing software like *Pegasus and the Phantom Riders* (inspired by arcade game *Joust*), *Interbank Incident,* and *Pan* (music composition), among several others.

As Figge recalled: "The Radio Shack relationship was project-by-project, but there was always back-and-forth as the project went through the design and development stages. So, although the payment didn't happen until the product was sold, we worked with them to make sure we delivered a product they wanted. Some ideas for programs came from us, others came from Radio Shack."

Tom Dykema was in college when he came across the Color Computer. He was only 21 years old when he started his business, T&D Software, in his parents' basement, selling cassette- and disk-based software on a subscription basis. "I hired a programming genius down the road to help me." Along with his helper, Dykema wrote software at the rate of about four programs a week for the subscription service. By the third year, T&D was doing well and advertising in *THE RAINBOW* but hadn't yet attended a RAINBOWfest.

"Lonnie was always trying to talk me into going to the fests," Dykema recalls. When he finally did attend, he found people lined up at his booth to buy both his software and blank disks (Figure 4.8). "There were no Best Buy stores at the time, so people came to my booth to get this stuff."

Dykema's most memorable RAINBOWfest was in Princeton, New Jersey. "Lonnie would have a party on Saturday night for vendors, with free alcohol." At the hotel bar where the venue was happening, Dykema spotted Brooke Shields, the world-famous model and star of the 1980 film *The Blue Lagoon*. In a moment of amazing courage, Dykema gathered up the nerve to ask her to dance. He soon found himself on the dance floor with the statuesque model and movie star.

For years, T&D's full-page ads were prominently displayed in *THE RAINBOW*, sometimes taking up four pages in a single issue. When Tom Mix stopped advertising and selling his well-regarded CoCo software as the market softened, T&D made a deal to pick up the remaining stock for $1,000. Dykema immediately began selling the titles, like the *Donkey Kong* clone, *Donkey King*, on his own, as Mix transitioned to the PC side of the computer business, eventually retiring in 2007.

Tom Roginski was another vendor and prolific advertiser in *THE RAINBOW* who built a successful CoCo-based business, OWL–WARE, out of Mertztown, Pennsylvania. Roginski's company was well known for selling Color Computer-compatible floppy disk systems and hard drives, as well as specialized serial and parallel port boards. The company even supplied floppy drives and controllers for the International Correspondence Schools, which used CoCos for its courses.

"RAINBOWfests were very profitable for us," Roginski recalled. It was an opportunity to sell directly to the customer, and sales were brisk. At one RAINBOWfest in Princeton, New Jersey, Roginski recalls having to deal with over $30,000 in cash. When he got to the hotel room, he threw the money on the bed in a 20″ pile. In a moment of celebration, Roginski's son dove into the pile of cash on the bed. It seems that despite its metaphorical nature, in this case, there really was a "pot of gold" at the end of *THE RAINBOW* after all.

Truth be told, though, while many other companies and individuals were able to find pots of gold of their own, there was no question Tandy was always in control of each rainbow. It knew better than anyone how to reap maximum profits from its own products, led by a simple corporate directive that emphasized control over everything from the marketing message to supply chain costs. It was emphasis on the latter, however, that would occasionally get Tandy into trouble.

5

Double Trouble

"Cost, cost, cost!" was the Tandy mantra, and one that Dale Chatham knew well—keep costs down and find ways to build products cheaper. This directive applied to every facet of work. Managers were asked to do more with less and to encourage their engineers to periodically review designs to determine if the company could take advantage of less expensive components or some other cost-saving measure.

Though perhaps not quite to Apple-esque levels, Tandy was still always famously focused on keeping profit margins high. For Tandy executives, the math was simple: the more a product sold, the more of an opportunity existed for cost reductions. The Color Computer was clearly a target for these cost reductions precisely because it was selling well in Radio Shack stores. Even just 5% shaved off of the manufacturing cost of a million-unit seller could represent $50,000 in savings.

One interesting outcome of this idea of leveraging low-cost, high-margin components was a product known as the MC-10, or Micro Color Computer (Figure 5.1). As the name implied, the MC-10 was a cut-down, scaled back variation on the Color Computer placed in a dramatically smaller case, often likened to the size of a hardcover book. It was released with little fanfare in November 1983 at a bargain price of just $119.95 for the base unit, which featured 4K of RAM.

The MC-10 retained the CoCo-style chiclet keyboard, albeit in a smaller, even less usable form factor, and the same 6847 VDG graphics, but this time paired

Figure 5.1

The TRS-80 Micro Color Computer, shown with one of its cassette tapes for scale.

with a slightly more limited Motorola 6803 microprocessor. According to Mark Siegel, Tandy wanted a product that was competitive with the Timex Sinclair 1000. "It was a secret R&D project," says Siegel, who ended up designing one of the few available programs for the machine, *Pinball Wizard* (1983), which, like all MC-10 software, was supplied on cassette.

Unfortunately for Tandy, time was not on its side with the MC-10. The idea of a limited function but low-cost computer first took off with the 1980 release of the British Sinclair ZX80, a tiny white computer with a completely flat, blue membrane keyboard. Base memory was just 1K and it could only output a black-and-white picture with no sound, but its saving grace was that it was the first prebuilt computer available for under $200 in the United States (under $150 if bought in kit form). This was followed about a year later by the even more successful ZX81, which, starting in July 1982, saw wider release in the United States under Timex as the Sinclair 1000. Though only slightly more capable than its predecessor, the Sinclair 1000, which featured 2K of RAM, was tantalizingly priced at less than $100. This aggressive pricing helped it sell well over a half-million units in a little over five months.

Unfortunately for all the new owners of the Timex Sinclair 1000, it was not much of a computer, even at its bargain basement price. To add insult to injury, Texas Instruments with its 16K TI-99/4A and Commodore with its 5K VIC-20 were locked in a winner-take-all competition for who could drop the price fastest on their far more capable computers, relegating the remaining stock of Sinclair 1000's to bulk bargain bins at $10 each.

In concert with The Great Videogame Crash, which began in late 1983 and continued through early 1985, the computer market was seeing its own contraction. If your low-end computer platform did not see release or establish a foothold prior to 1983, public interest would prove minimal. Though many companies tried, the few survivors in this segment of the computer market were already set, particularly when Commodore effectively shut any remaining competitive windows with its successive, aggressive price drops on its powerful 64K Commodore 64 computer following its late 1982 launch.

Nevertheless, even without the challenges of a market correction, by the time the MC-10 was made available in Radio Shack stores, there was little reason to own one. The MC-10's small footprint offered no advantage, since it still needed to be plugged into a TV. The keyboard was difficult to use and there were few expansion options other than printers and a 16K RAM module, which was required to access all of the VDG's video modes and added another $49.95 to the price. This left it just $70 shy of the far more versatile 16K base model of the Color Computer 2, complete with a real keyboard and a robust selection of software.

Whatever the reasons, in the end MC-10 sales clearly did not meet Tandy's expectations. Dale Chatham, who worked on the engineering side of the computer, recalls: "I remember this being a pretty simple and quick response to the Sinclair ZX81. We didn't put a lot of work into the development of the product, and I guess it showed in the lack of success." The miniature variation of the Color Computer was discontinued just one year after launch. It was not all bad news for Tandy, however, as the MC-10's demise left more time to focus on the value proposition of the Color Computer 2, whose engineering champion was one John Prickett.

Prickett's career at Tandy started in February 1980, but his interest in electronics began much earlier. An article by Don Lancaster in the September 1973 issue of *Radio-Electronics* magazine caught his eye: "TV TYPEWRITER." In the article, Lancaster provided the basic details to create an innovative character generator for a TV set that could form the basis of a computer terminal, educational toy, or display device. Intrigued, Prickett successfully built the device and became hooked on electronics from that point on. After a stint in the Air Force, Prickett left the service and spent subsequent years dabbling further in hobby electronics and TV set repair.

In 1978, Prickett purchased a SWTPC 6800 Computer System in kit form for $399 from the Computer Port store in Arlington, Texas, and stuffed it with integrated circuits that he purchased from various sources. For a video screen, he used a JCPenney brand 9″ black and white TV that a customer had brought in for repairs. "Penneys wanted more for the tuner than the set was worth," recalls Prickett, "so the customer abandoned the set rather than pay the $7.50 checkout fee."

To get around the tuner issue, Prickett simply tapped into the television's video amp, turning it into a perfect monitor. The keyboard turned out to be the most expensive part of his new terminal—he found one made by George Risk Industries for about $65.

Pursuing his master's in music theory while attending Southwest Baptist Theological Seminary, Prickett wrote his 100-plus page thesis "A Set Theory

Analysis of Messiaen's 'La Nativite Du Seigneur,'" on his custom SWTPC micro-computer using a simple line editor. By that point, he had acquired a Smoke Signal Broadcasting floppy disk controller with a Shugart SA400 single-sided 35 track single-density drive. "One disk held 80K," Prickett remembers. In order to print his thesis from his computer, he built an interface for his IBM Selectric typewriter.

His next addition was Smoke Signal Broadcasting's "The Chieftan," a 6809-based computer. The graphics and memory cards were made by GIMIX out of Illinois, while the case, power supply, and many of the other peripherals were still from SWTPC. Before long, Prickett began to dabble in 6809 assembly language. As he studied the architecture of the processor, he became enamored with the powerful 8-bit CPU.

After graduating in 1979, Prickett applied for and was given a job at Tandy in February 1980. His assignment: teaching Tandy's technicians how to repair Model I and II computers, even traveling as far as Belgium to do so. With an educational background, Prickett's job as an instructor was a natural, but he was much more eager to work his way into engineering.

Back in the halls of the Tandy Towers, Prickett got wind of the Color Computer project, and sought Dale Chatham to learn more about the 6809-based home computer. Chatham was pleased in the interest that Prickett showed in the project, and was impressed with his desire to join engineering. Encouraged that his favorite microprocessor was going to be a computer project at Tandy, Prickett began looking for ways to prove to Tandy that he had what it took to meet the challenges of being an engineer. It would not take long for an opportunity to arise.

In 1981, Prickett's son was born 53 days premature, resulting in a 28-day hospital stay and extended bills. Tandy's insurance provider, unwilling to pay the full amount, left the family in a lurch. Facing the extra out-of-pocket expense, Prickett was looking at a large, looming hospital bill. In order to meet his obligation, he took a second job on weekends, babysitting the radio transmitters at 1540 AM, KTIA, a station that catered to Spanish programming. Working three shifts each on Saturday and Sunday, Prickett found himself mostly idle. Other than the occasional Federal Communications Commission (FCC)-mandated radio frequency (RF) measurement, there was little to do except wait for something to go wrong. Prickett's FCC measurements did have an interesting side effect, however. "The RF energy near the transmitters was so strong, it could light up a fluorescent light that you held in your hands."

With the extra time, Prickett began to toy with the idea of building an 80×24 video card for the Color Computer. The intent was to show the folks in Tandy System Design that he had the chops to work with their engineers. Starting with an MOS Technology 6545 video address generator, Prickett designed and hand-wired his card, and even located the hooks in BASIC so that he could write the necessary driver extensions to use the card in that environment.

Once completed, Prickett showed the working and completed card to the engineering folks in Tandy System Design. Bill Wilson, who headed the department, was impressed with both the card and the initiative that Prickett had taken, and decided to bring Prickett on board. "My first assignment was to finish the 'toaster,' a project started by Gerald Gaulke," Prickett remembers. The toaster was none other than that most useful of Color Computer accessories, the TRS-80 Multi-Pak Interface.

In the fall of 1982, Prickett got the major assignment he had been waiting for: designing the second incarnation of the Color Computer. This was to be an improved version of the original CoCo with the dual purpose of maintaining the same feature set, while lowering production costs. The opportunity itself came out of Tandy's near miss attempt at entering the console gaming market.

This was a time when home videogame consoles like the Atari 2600 VCS and Mattel Intellivision were still selling like hotcakes. According to Mark Siegel, Tandy's growing list of increasingly impressive Color Computer cartridges was a strength that Tandy wanted, particularly since the games market for the older TRS-80 computers was already drying up. Thus began the process of exploring a gaming console based on the Color Computer's core technology.

To hedge its bets, while Tandy continued down the path of pursuing the creation of a gaming system, it considered another option: buying a competitor's product and putting its own brand on it like it did with the TRS-80 Pocket Computer, which was actually made by Sharp. One possible path was completely taking over Mattel's Intellivision product line, a console that Tandy was already rebranding in its stores as the Tandyvision One. Siegel remembers: "The game machine was going to be in the MC-10 packaging with a cartridge slot on top. It would have used a membrane keyboard and 4K RAM. This was the backup plan if the purchase of Intellivision fell through." Siegel continues: "We almost did buy it from Mattel. When we demoed CoCo games compared to Intellivision games, it became obvious to management that we already had a superior platform for games. Tandy Electronics said they could meet the Intellivision pricing with a new cost reduction. I said that I could make the software work without the Microsoft royalty. At that point there was no reason to buy Intellivision."

With any thought of purchasing Intellivision now discarded, the Color Computer gaming console was reconsidered. However, it became apparent that there was little advantage to creating a gaming system over a conventional home computer. "When we were done, the only difference between a Coco 2 and a game machine was the keyboard and RAM, which was offset by the cost of joysticks on the game machine. So there was no point in doing a game machine. We should just do the CoCo 2."

The course was set on a new Color Computer, and Prickett made plans to make it happen. He would first tackle the consolidation of components. Doing so would drive down the part count as well as the size of the motherboard, thereby achieving the cost and margin objectives that Tandy had set out for its hot-selling home computer.

One task was to combine the Motorola MC1372, a video mixer circuit, with the RF modulator. Prickett got Tandy Japan to do the consolidation and realize the implementation. With the lessons learned from the pseudocolor fiasco in the low-cost version of the Color Computer that Paul Schreiber had worked on, Prickett set out to inject noise into the circuit so that the monochrome 256 × 192 resolution graphics screen could exhibit the false colors that game programmers had taken such great advantage of.

Unlike Schreiber's pseudocolor fix, however, Prickett had to take a different approach, given the consolidation of the 1372 within the RF modulator. With Prickett's television repair experience, he had intimate knowledge of how TV signals worked, specifically those signals broadcast throughout North America using the National Television System Committee (NTSC) standard.

With a scope and logic analyzer, Prickett began to determine just how the 6847VDG and MC1372 chip generated color burst. He explains the process: "The way the Motorola chips work is that in their color modes, when they get to the back porch of the horizontal sync pulse, they pull either the Chroma A or Chroma B and unbalance the modulator that was in the 1372. I needed to trigger something off of the horizontal sync pulse and use a transistor to pull one of those chroma lines, and just randomly try them out until I got either red or blue. I was also able to detect off to the PIA [peripheral interface adapter] a signal that switched the modes in the 6847, and inject the color burst for that mode only."

A 555 timer circuit was used to inject the color burst at just the right time, but getting the right color was still hit or miss. Depending on how the CoCo 2 powered up, one would either see orange-red or blue as the primary color, a "feature" that can readily be seen on CoCo 2's today. Prickett used Tandy's text and graphics adventure, *The Sands of Egypt* (Datasoft, 1982), which took advantage of the pseudocolor mode, to tweak the feature.

The experience left Prickett to conclude that NTSC actually stood for Never The Same Color.

When Prickett told Mark Siegel that he successfully solved the pseudocolor issue on the Color Computer 2, Siegel was skeptical. "Since Paul Schreiber's pseudocolor solution for the Color Computer didn't work on all television sets, he [Siegel] didn't believe mine would work either. I bet him a steak dinner that it would." In the end, Prickett's solution was indeed proven to work on all television sets.

The steak was delicious.

The other key part of consolidation had to do with the analog circuitry responsible for providing voltage to the cassette and RS-232 ports. On the Color Computer, multiple components performed the functions. For the CoCo 2, Prickett began working with Motorola to consolidate that functionality into two chips: the Supply And Level Translator, aka, the SALT chip (SC77527P); and the Digital to Analog Converter, aka, the DAC chip (SC77526P).

At Motorola, engineer Dave Elmo took on the design of the SALT, while Ira Miller focused his efforts on the DAC. With the specification of the chips defined and put into place, Motorola began constructing breadboards of each chip. The larger scale prototypes were assembled so that a pod would easily plug into the socket on the Color Computer 2 prototype, giving the engineers a way to "bring up" the board. Over the course of four trips to Motorola's semiconductor facility in Tempe, Arizona, from Fort Worth, Texas, the engineering team of Elmo, Miller, and Prickett managed to get the prototypes working.

On one of the later trips, Prickett recalls carrying two prototype CoCo 2 boards with him. "We were going to test the SALT and DAC prototypes, and if everything worked, I was going to carry them back with me to Fort Worth."

Prickett arrived at Motorola and the engineers began to connect the breadboard SALT and DAC prototypes to the CoCo 2, itself just a prototype. The moment came to turn on power and see what would happen. To everyone's relief, the familiar green screen with black text appeared—along with something else unexpected: an odd green-colored vertical bar that filled about 1" of the left edge of the 19" television. Head scratching commenced as the engineers wondered what was going on with the video signal.

"Since we had two CoCo 2 prototype boards, I just hooked the other one in place and turned it on to see if it would behave the same way." It did not. The second prototype board showed the expected video without the strange green bar. Something was wrong with the first CoCo 2 prototype.

After studying both boards, Prickett pointed out in jest, "This one works because it has a Fairchild 555; the other one has that piece of shit Motorola 555." With further diagnosis, it turned out that Prickett's humored observation was indeed correct. The Motorola 555 timer circuit was firing 1.5 times per cycle instead of 1 time per cycle, leading to the strange artifact on the video. The problem fortuitously led to the discovery of a production issue that affected large runs of the part.

As the SALT's design firmed up into a 16-pin dual-inline pin package, Prickett discovered that the chip's transistor would only pull 70 milliamps of current from the 5 volt coil in the onboard cassette relay. This limited the choice of relay that could be used. Prickett's research led him to Aeromat, an electronics supplier that made a relay that could handle the 70 milliamp requirement. One problem remained, however. The relay had more contacts than was needed, and Prickett saw a chance to negotiate on price. He proposed to the Aeromat representative that a relay with the same current draw constraints be manufactured, but instead with a single contact relay. The representative balked at the suggestion and went on to explain that such a thing could only be done in very large quantities.

In a moment that succinctly illustrated Tandy's enormous buying power, Prickett chuckled to himself and replied, "We'll need 200,000 a year for several years." The Aeromat company man quickly changed his tune and indicated that there would be no problem meeting those requirements.

Other improvements over the original Color Computer were put in place. Cassette volume sensitivity was too high on the Color Computer, which caused problems in loading programs if levels were not set exactly correctly. Prickett made that a nonissue on the Color Computer 2. The keyboard, which was procured from Tandy's subsidiary in Japan, A&A, would be an improvement over the chiclet keyboard of the Color Computer. Real, closely spaced keys were used and the "travel" distance increased, which gave a more natural feel while typing. The key profile was still low to the board, giving the appearance of a melted look and its unfortunate nickname. Another cost-reduction feat was to condense the CoCo 2's printed circuit board to a size where Tandy could yield six boards per panel instead of the two boards per panel of Schreiber's Color Computer cost and reduction efforts.

As the CoCo 2 approached final production, Prickett completed the task of writing the computer's service manual. When the first Color Computer 2 rolled off the assembly line in Fort Worth, Kenji Nishakawa presented serial #0000001 to Prickett, like a doctor giving a newborn baby to its father. The newly minted computer in all its bright white glory sported the mouthful of a label, "Radio Shack TRS-80 Color Computer 2."

Word was out that new Color Computers were going to be released. *THE RAINBOW* magazine's September 1983 issue scooped the story, its cover showing the 64K Color Computer in its fresh fawn gray case bursting through the front cover. The following month, Falsoft put on their second RAINBOWfest ever in Fort Worth, Texas, right at the doorstep of the CoCo's home. Tandy provided some support to the event by having a booth there and even offering tours of the factory where the Color Computer was made. Considered a flop by some, the Fort Worth RAINBOWfest would be the first and last of its kind in the Southwest. However, the event showed Lonnie Falk that there was still positive interest in the little machine.

Among the hall walkers was the CoCo 2's own John Prickett. Shunning celebrity status that he could have easily and rapidly attained had he announced his presence, he instead chose to simply visit the many booths of wares, without bothering to tell anyone who he was.

On Friday, September 9, 1983, Tandy officially announced the Color Computer 2. On cue, the 1984 Radio Shack catalog appeared in company and dealer stores as well as mailboxes around the United States.

Contained in the catalog's pages was the new Color Computer 2 in two configurations: a 16K Standard BASIC configuration for $239.95, and a 16K Extended Color BASIC edition for $319.95. The original Color Computer also received an update thanks to Schreiber's cost-reduction efforts. Gone was the dull gray case. The CoCo had graduated to a full 64K in a fawn gray-colored case for $399.95 (Figures 5.2 and 5.3). The new Multi-Pak also made its debut for $179.95.

By this time, the retail price of its main competition, the Commodore 64, had dropped to well below $300 with no signs of stopping, but Tandy was still banking on the massive reach and unified marketing message enabled by its catalogs and stores to make up the difference. As David H. Ahl so succinctly put it in the March 1984 issue of *Creative Computing*, "One reason that Radio Shack has been so successful—even with less than competitively priced products—is that product is on the shelves to some 6,000 stores. That's hard to beat."

Production of the Color Computer 2 proceeded at a fast clip in the Fort Worth factory in order to get the product to stores in time for the Christmas holiday shopping season. Prior to the wave soldering process that welded chips into place, the bare motherboards would be stuffed with parts. After the soldering, the boards were then baked in ovens in order to harden them. The hardened boards were then examined, tested, and paired with cases, keyboards, and power supplies, then off to quality control for additional checks and testing.

As Jim Reed from Falsoft recalls, he and Lonnie Falk visited Tandy's headquarters at Fort Worth a number of times, and on at least one occasion toured

Figure 5.2

The new Color Computer 2 as it appeared in the 1984 Radio Shack Catalog. (Courtesy of www.RadioShackCatalogs.com.)

the factory. Most who worked there were Laotian, and hardly anyone there could speak English. "It was like walking into an Asian factory right in the middle of Texas!" said Jim.

All of that changed in 1985. Dale Chatham left the Color Computer and moved over to the Tandy 1000 group to design the video ASIC for that upcoming system. He recalls the decision to move the Color Computer manufacturing overseas. "The Fort Worth factory didn't have the bandwidth to support both the Tandy 1000 and the Color Computer." Tandy 1000's would be made in Fort Worth, while the Color Computer 2 would be produced in a factory in Korea for the remainder of its life. Dale continued, "Overseas production of the Color Computer 2 started

Figure 5.3

The 64K Color Computer. (Courtesy of www.RadioShackCatalogs.com.)

after I made a trip to Tokyo, Japan, along with several Tandy and Radio Shack vice presidents. Tandy/Radio Shack had acquired a small engineering group that worked just outside of Tokyo, and they used that group to support the overseas factory that built the Color Computers. Needless to say, I felt pretty intimidated being the only lowly engineer on a trip with all of the corporate VPs."

Chatham relates a humorous story on the trip to Japan. "I wound up giving Americans a reputation for not liking seafood. The trip lasted about a week, and one evening I was going with all of the VPs to a seafood restaurant. I do not like exotic seafood, and the restaurant that John Patterson and the others picked served all types of seafood. All it took was walking in the door, and the smell—along with being nervous about being out with the VPs—was enough to make me sick. I had to leave and go back to the hotel in a cab. Also, we had lunch several times at the work location for the engineering group. On one or more occasions, they served a seafood soup, and I am sure that they took note of the fact that I ate only a small amount of the soup. Also, one time, they took us to an American restaurant—*Denny's* I think—and I ordered my favorite: a hamburger and fries. Again, I am sure they saw that I ate every bite of that. So that is how I started the rumor that Americans don't like seafood. I was told that for several years after my visit, American visitors to the plant were never served seafood."

With the CoCo 2 complete and finally put to bed, Prickett turned to the Color Computer's line of peripherals for rework and cost-reduction opportunities. One such peripheral was Tandy's FD-500 disk controller. Cutting the size of the elongated cartridge in half would not only reduce printed circuit board costs, but it would also save on the plastic material that was used to make the case that enclosed the board. Prickett reworked the disk controller, which Tandy dubbed the FD-501, around the end of 1983.

He next focused his attention to his first love: music. Music was an appealing idea for a Color Computer add-on product. In addition, Tandy's Mark Siegel wanted to give the CoCo the ability to talk more easily, too. Inspired by a Navy study on text-to-speech, Siegel went to work convincing upper management that the idea of a talking, singing Color Computer was a good idea. With their blessing, he began defining the requirements for the product, and Prickett went to work on the design.

Prickett's first task was to choose a chipset and was aware that Texas Instruments had a text-to-speech silicon solution using a DSP with a number of ROM chips. While the solution was robust, the cost was not insignificant at $30. Being ever-sensitive on cost, Bill Wilson, Tandy System Design's director of engineering, prodded Prickett to find a cheaper alternative.

With some additional searching, Prickett stumbled upon another solution: General Instruments offered a two-chip solution for speech and music. Both the text-to-speech chip and the AY-3-8913 sound chip cost $6, exactly one-fifth the cost of the Texas Instruments solution. Wilson gave the go-ahead to use the chips for the design. In addition to the cost of the printed circuit board, he gave Prickett a budget of just 50 cents to spend on all the remaining electronic components, including any required transistors, resistors, and capacitors.

Along with an Appliance/Light Controller, the X-Pad, and the Electronic Book, the Speech/Sound Program Pak debuted in the fall of 1984 and appeared in the 1985 Radio Shack Catalog for $99.95. In Tandy's often successful quest to boost gross margins, the estimated build cost for the talking cartridge came in at just under $25.

Although the CoCo 2 represented a sleeker, more cost optimized Color Computer, it still offered no real additional features over the Color Computer that preceded it. Tandy engineer Dale Chatham knew that what CoCo users wanted was more sound, more graphics, more memory, and more speed. The balance between features and cost was always tilted toward the latter, but it did not stop engineering from trying new things. That is what spurred the legendary Deluxe Color Computer, a grand experiment in making a better, more powerful CoCo.

As the father of the Color Computer, Chatham was well aware of where the machine's weaknesses lay. So, along with Mark Siegel, Chatham worked to define an impressive set of features that would elevate the Color Computer to the level of a more serious system. To start, there would be an enhanced keyboard with real CTRL and ALT keys, arrow keys that were grouped together, and function keys, F1 and F2, added to the existing CoCo keyboard footprint. Gone would be the chiclet or flattened keyboards. In its place, raised keys that had more travel.

The 6-bit DAC, which was responsible for sound on the original Color Computer, would keep its role as a joystick input comparator but would defer its sound responsibilities to General Instruments' AY-3-8913, the same chip found in the Speech/Sound Pak. However, the chip would be directly integrated into the motherboard, providing applications with universal access to a rich set of audio features. In addition to the 4-pin software serial "bit banger" port, there would be a "real" 9-pin serial port with a 6551 Asynchronous Communications Interface Adapter (ACIA). The joystick port would gain an extra input pin, providing support for joysticks with two action buttons.

The Motorola 6847 Video Display Generator (VDG) chip, which had suffered the indignity of reverse video characters to compensate for the lack of true lowercase, would be replaced with a new and improved 6847, which would provide true lowercase with descenders, in addition to compatibility with the preexisting reverse video characters.

Plans were in place to have the full 64K of RAM accessible from BASIC; 32K would be used for standard memory, while the other half would be accessible as a high speed RAM disk. A 32K ROM would house both Microsoft's Standard and Extended Color BASIC ROMs onboard, as well as Disk BASIC, with extensions to support the RAM disk.

The RF modulator would remain on board for video hookup to a television, but the Deluxe Color Computer would also sport composite audio and video, allowing it to interface to the better monitors of the time. Finally, the board itself would fit inside the same case as the original Color Computer.

Design commenced and eventually prototypes rolled off the assembly line. Several assembled systems were sent to developers, including game programmer Greg Zumwalt, who was tasked with adapting the ROMs. Systems engineer Mark Hawkins of Microware Systems Corporation in Des Moines, who worked on OS-9 for the Color Computer, brought that operating system over to the Deluxe Color Computer. "It ran OS-9 beautifully," recalls Mark Siegel.

Excitement was building inside and outside of the Tandy Towers. Although the company was usually secretive, opting to wait for its yearly catalog to announce the availability of products, word leaked to the larger CoCo community about a new, more powerful Color Computer. The rumor mill began to churn. Would it have a 40- or 80-column screen? What about more memory, like 128K? Or 320 × 200 high resolution graphics? Speculation was rampant about what Tandy might deliver.

But all was not well with the Deluxe Color Computer. As the summer of 1983 approached, Chatham and Siegel faced the hard fact that the promising system was running into cost overrun issues. The price of the General Instruments sound chip alone was becoming a significant part of the cost of the computer. Added to that, Motorola could not guarantee enough production of chips that were common to both the CoCo 2, which was already selling, and the yet-to-be-produced Deluxe Color Computer.

Tandy higher-ups decided that given the choice, the Deluxe Color Computer would have to go. Siegel remembers how he felt when he got the word of the cancelation: "I was devastated. We had put a lot of work into it."

Chatham recalls: "The Deluxe Color Computer was a temporary flirtation with adding features requested by outside software developers. However, that version was quickly discarded when accurate estimates of the manufactured cost of the product were available."

He continues: "The timeline for all of our significant products at Tandy was to be available for sale in Radio Shack when the fall catalog came out. Also, the timing for the catalog and the products was to have volume quantities ready for sale before Christmas … the concept for the Deluxe version was considered first, but was discarded relatively quickly in favor of [solely] the Color Computer 2."

Stacks of motherboards sat at the Fort Worth factory ready to be stuffed with parts when they, along with the cases, were unceremoniously destroyed. The keyboards, however, were salvaged. According to Mark Siegel, some were shipped to Radio Shack stores and sold in surplus bins for as little as $5, but most were repurposed for the Tandy Color Computer 3, which would arrive on the scene in 1986.

Even after the cancelation of the Deluxe Color Computer, several tantalizing clues were left, quite accidentally, in the Color Computer 2's "Getting Starting With Extended Color BASIC" manual. There, on page 18, a sidebar reference states: "If you have a Deluxe Color Computer, your computer can understand commands in 'reverse' or 'lowercase' type. See *Introducing Your Deluxe Color Computer* to learn how to get in the upper/lower case mode." In several sidebars of the book, references were made to the "Deluxe Color Computer." A similarly worded note appears on page 158.

Yet another sidebar note, on page 53, reads: "If you have a Deluxe Color Computer, EDIT will not work for you. You have a better way of editing program lines—the ALT key. The ALT key is described in *Introducing Your Deluxe Color Computer*." Later on page 199, the final tantalizing sidebar note reads: "If you have a Deluxe Color Computer, use the CTRL key rather than ↑." These references remained in the teaching books that were shipped with CoCo 2s as late as 1985, some two years after the Deluxe Color Computer was canceled.

Today, there is little evidence left of the Color Computer that almost made it to market. Only two Deluxe Color Computers are known to exist. Both are operational and presently in the hands of a collector (Figure 5.4).

Color Computer sales outside of the United States were handled by Tandy owned subsidiary InterTan. CoCos were sold in Canada (whose "Colour Computer 2" spelling on the product boxes were a telltale sign), as well as in

Figure 5.4

The Deluxe Color Computer never made it to market. Today, only a few remain in the hands of a collector.

Europe and Australia. For those markets, the CoCo 2 motherboard had to be slightly redesigned to accommodate the 50 Hertz refresh rate of PAL-encoded television sets, but aside from that, they were identical to their U.S. counterparts. Color BASIC and Extended BASIC manuals, which were well written and suitable for people of all ages wanting to learn to program BASIC, were translated into other languages, including Dutch, German, and French.

The CoCo platform under Tandy enjoyed great success in the U.S. and Canadian markets, but competitors were waiting in the wings. The "attack of the clones" began in 1982 with the announcement of the Dragon 32, a home computer made available to customers in the United Kingdom from British computer company Dragon Data. Although its case and keyboard were markedly different from that of the Color Computer, internally, the Dragon used the same 6809 microprocessor, 6883 SAM chip, and 6847 VDG chipset.

Unlike the CoCo, which came with only a BASIC manual and an RF switch box and cable, the Dragon 32 featured those items, plus a suite of cassette-based software, including a word processor, mailing list utility, and a simple spreadsheet and database. Compatibility with the Color Computer was solid for many of the programs written in BASIC, requiring mostly minor changes to work, but machine language programs required additional effort. There were other key differences, including how the full stroke keyboard was wired, and the number of available ports, which included monitor output. For the potential CoCo enthusiast, the Dragon held great interest, but its lack of out-of-the-box compatibility would clearly challenge wider initial adoption until its software caught up.

Other CoCo clones soon followed. In Taiwan, Sampo announced the Sampo Color Computer in the November 1982 issue of *BYTE* magazine. Though little is known about its features, it too was based on the 6809 microprocessor and featured Microsoft's Extended BASIC. As far away as Yugoslavia, a repackaged Color Computer, known as the Misedo 85, was targeted to schools in that country (Figure 5.5).

In Brazil, there was the CODIMEX-6809 (Figure 5.6), the MX1600 from Dynacom, the Color 64 from LZ Equipamentos Eletrônicosas, as well as the CP-400, a Color Computer clone from Proligica. Although it had a different exterior than that of the CoCo (copied from, of all things, the Timex Sinclair 2068), the CP-400 used the same Motorola chips, making it closely compatible. Even Microsoft's Color BASIC was adapted to the machines by their respective manufacturers. Juan Carlos Castro of Rio de Janeiro, who worked for LZ, which sold the Color 64 explains: "Back then, by Brazilian law, foreign companies didn't have any copyright claim to hardware designs or software, so we freely copied, patched, and enhanced Color BASIC, Extended Color BASIC, and Disk Extended Color BASIC. Basically, every Brazilian clone machine of the time was a bootleg—but legal."

Public schools in Mexico were introduced to their own version of the Color Computer through a project called COEEBA-SEP. The project was started by the Latin American Institute of Educative Communication (ILCE) along with the Public Education Secretary (SEP). Known as the microSEP 1600, the computer's

Figure 5.5

The Misedo 85, a Color Computer 2 clone made available in former Yugoslavia. (Courtesy of PC Press, http://pc.pcpress.rs/pcmuzej.)

Figure 5.6

A 1983 ad for the CODIMEX-6809, one of several Color Computer clones from Brazil.

motherboard closely resembled that of an original Color Computer, except the RF modulator was replaced with composite video (Figure 5.7). Educational programs were supplied on elongated cartridges, and later, diskettes.

Although the CoCo's clone competition started outside of the United States, that would change in 1983. Tano Corporation of New Orleans, Louisiana, was a multimillion dollar corporation that made its mark providing electronic systems to companies in the oil and gas industry. But that industry had taken a hard hit

Figure 5.7

The microSEP, a rebadged CoCo 3 for the educational market in Mexico.

CoCo: The Colorful History of Tandy's Underdog Computer

in 1982. Looking to diversify, Tano began to search for other opportunities to continue its impressive growth. Drawing on its background in electronic industrial control systems, aka, SCADA and energy management systems, Tano felt confident that it could apply their computer-building expertise into the profitable and fast-growing home computer market.

After researching the current offerings, Tano initially decided on an Apple II work-a-like from a Dutch firm that would be manufactured in Korea, but backed away, fearing potential litigation by Apple Computer. As an alternative, it chose Dragon Data's Dragon 64, which featured twice the memory (64K) and the addition of an RS-232 serial port over its Dragon 32 predecessor. With a distribution deal signed between Dragon Data and Tano, the attack of the clones had arrived in Tandy's own backyard.

Tano's Dragon went into production in late 1983, and an advertising blitz accompanied the announcement, creating a significant buzz (Figure 5.8). Ads appeared in CoCo magazines, including *THE RAINBOW*, touting the upcoming Dragon by Tano as an alternative to the CoCo in the United States (Figure 5.9). In addition, *Dragon User* magazine out of the United Kingdom began to add coverage on the new sister machine to its pages.

Storm clouds were on the horizon, however. During 1983, The Great Videogame Crash began to take its toll, and in its wake were big losses for computer companies like Atari, Texas Instruments, and Osborne. It was in this environment that Tano announced two consecutive quarters of losses. Ironically, its effort to diversify outside of the volatile oil and gas business placed the company in the midst of yet another industry shakeup. Cutting its losses, owner Jim Riess and his fellow management team made the unfortunate but necessary decision to abandon the Dragon, and the home computer market itself, in 1984. Remaining stock of the

Figure 5.8

The U.S. version of the Tano Dragon was manufactured by Tano Corporation in New Orleans, Louisiana.

This, quite literally, is the color computer America has been waiting for. One of the best sellers in the United Kingdom, the Dragon will soon be manufactured by TANO to serve American consumers who want a serious, affordable computer; one that has proven itself at providing educational and home management applications as well as fun and games. To meet this challenge, the Dragon was born. With a standard 64K of RAM. A professional typewriter-style keyboard guaranteed for 20 million key operations. And an impressive array of options which include disk controller and drive, a printer, audio cassette recorder, a modem (RS-232 serial I/O), joysticks, game cartridges and a free BASIC training manual.

DRAGON! THE COLOR COMPUTER YOU'VE BEEN WAITING FOR.

And full editing features allow you to insert, delete or change at will. Oh yes — don't forget the colors! The Dragon features nine; with five different resolutions from 512 points of text (16x32) to 49,152 points (256x192) at high resolution. And you can view these amazing phenomena through either your composite video color monitor or VHF TV. So goes the Dragon's story. If you'd like to know more, just mail the coupon or call George Merchant (our Director of Marketing) toll free at 1-800-327-7671. Software developers and dealer inquiries are welcome. The Dragon is destined to become legend as America discovers its great performance is no myth.

Using the new 6809E Microprocessor (a great advance on the original 6502 still used by our competition), the Dragon brings advanced computer functions well within your reach. And priced below $400, it's anything but expensive.

But these aren't the only points of difference with which our Dragon roars. Unlike most units, the Dragon gives Extended Microsoft™* Color BASIC as its standard language while the competition is still stuck in Microsoft™* BASIC training. The Dragon's advanced graphics features include set, line, draw, circle, paint, print @ and print using. Of course the Dragon also features advanced sound capabilities.

*Microsoft™ is a registered trademark of Microsoft Corp.

Please send me more information on the Dragon.
NAME _____
COMPANY _____
ADDRESS _____
CITY _____ STATE _____
ZIP _____ PHONE _____

4301 Poche Court West
New Orleans, LA 70129

TANO
MICROCOMPUTER
PRODUCTS CORP

Figure 5.9

Tano's ad in the September 1983 issue of *THE RAINBOW*. Despite a promising marketing campaign, the machine would soon fail in the marketplace.

Tano Dragon was liquidated and sold to a surplus company on the West Coast, California Digital, which continued to carry inventory well into the 2000s.

If Tandy had any concern about the CoCo's clones, there's no evidence to suggest it. Tandy's reaction to the existence of the clones seems to have been muted. According to Mark Siegel, Tandy knew of the blatant copying of its hardware and software in other countries, but in most cases did not consider them a threat. Attorney Gary V. Pack, who worked as counsel for Tandy Corporation from 1979 to 1985, does not recall any specific litigation regarding the Color Computer clones either.

Despite The Great Videogame Crash, the CoCo 2 continued to sell well during the Christmas season, so much so that it had an appreciable, positive influence on Tandy's bottom line, which was no surprise considering how heavily it was advertised in everything from the company's catalogs to free comic books (Figure 5.10). John Prickett recalls being present at a meeting in the Tandy Towers where Radio Shack president Bernie Appel was venting about the depressed price of Tandy's stock. At one point during his admonishment, he picked up a CoCo 2 sitting on the conference table, lifted it up in the air, then loudly proclaimed:

"And if it wasn't for *this*, our stock would be even lower!"

Figure 5.10

A page from the March 1985 comic "Alec and Shanna Starring in the Computers that said NO to Drugs!" which was part of the popular Archie Comic Publications' produced *Tandy Computer Whiz Kids* series. These free comics gave kids everywhere an excuse to visit their local Radio Shack store and dream about their own, preferably CoCo-powered, computer adventures.

6

Silicorn Valley

The dark, rich alluvial soil that covers the fertile, rolling hills of central Iowa like a thick blanket is known the world over for producing an abundance of corn, soybeans, and other crops. This Midwestern state is hearth and home to one of America's most revered institutions: the farm. During the summer and fall months, acres upon acres of green and growing corn stalks paint the horizon as far as the eye can see. In late autumn, an army of green John Deere tractors and combines can be seen making their well-ordered rounds in the fields, harvesting their golden bounty.

Iowa is also the land of the swine. Factory farms dot the rural landscape, rolling out pork bellies and hams bound for grocery stores and supermarkets across the country.

This picturesque, agrarian state of Iowa was a world away from the technological mecca of Silicon Valley in California, but that did not keep the Hawkeye State from having its own hi-tech darling, tucked away in Clive, a suburb of the state's capital, Des Moines. All one had to do was to drive north on 114th Street, past the headquarters of the National Pork Board, to find a nondescript, one-story complex of buildings. This was the home of Microware Systems Corporation, a company whose product would come to play a prominent role in Tandy's Color Computer (Figure 6.1).

Figure 6.1

Microware Systems Corporation's headquarters on 114th Street in Clive, Iowa (circa 1985).

Before Ken Kaplan founded Microware along with Larry Crane and Robert Doggett, he was a student at Drake University in the early 1970s majoring in computer science and focusing his studies in the promising field of artificial intelligence. Kaplan's interest in the thinking computer led him to research and design an odd-looking hardware device called a "turtle," which was a small educational robot that could be programmed to move in tight, geometric patterns. The turtle concept was closely associated with Logo, an educational programming language created in 1967 that would peak in popularity in the 1980s, with versions for most of the era's personal computers, including the CoCo, with programs like the *TRS-80 Color Logo* (1982) and *D.L. Logo* (1985).

In searching for a microprocessor to run the turtle, Kaplan turned to Motorola, who had announced its 6800 microprocessor in 1974. After locating a sales representative and requesting samples of the processor, Kaplan lucked out and ended up with a Motorola-donated development board with 1K of RAM, an early experimental version of the 6800 chip, and a cross-compiler to create the necessary software. Using the university's CDC-6600 computer, Kaplan wrote the code needed to get his turtle device working and connected to another system, which would act as the real-time control mechanism.

The turtle project turned out to be a success for the young student, and Motorola took notice, helping Kaplan to garner a handful of hardware and software development consulting gigs. During the course of development of the turtle, Kaplan used Motorola's supplied tools, including its MkBUG monitor program. Working with that product gave him the idea to write a simple real-time kernel,

CoCo: The Colorful History of Tandy's Underdog Computer

dubbed RT/68, to support input/output (I/O) devices like cassette recorders and teletypes. Seeing the potential for such software, Kaplan decided to market RT/68 and form a company around the product. And so in 1977, Microware Systems Corporation was born of humble roots, in slum housing near Drake University.

As a newly minted corporation with few resources, it was decided that the best move for Microware was to place an ad in one of the most popular monthly computing magazines, *Byte*. It was the right move, as Kaplan recalled, because shortly thereafter, "the checks started coming in the mail." Fueling sales of RT/68 was the advent of 6800 systems from companies like SWTPC. The foundation for Microware's growth was set, and the company was soon moved to an apartment, then to a small building, and after that to a section of a warehouse next to a soybean plant in the northeast corner of Des Moines (Figure 6.2).

In those early days, the cash-strapped Microware leveraged its relationship with Motorola, who lent a hand to the young but burgeoning company by keeping an inventory of expensive ROM chips, which housed Microware's software. As orders came in, Motorola would ship out the ROM chips to Microware, who would then send them to paid customers.

That same year, in 1977, Motorola began development of its next-generation microprocessor, the 6809. Terry Ritter, a young electrical engineer who worked for Motorola in Austin, Texas, contacted Microware's Kaplan and sold him on the idea of supporting the up and coming microprocessor with a language that would take full advantage of the features of the chip. Motorola was interested in promoting the BASIC language for its new microprocessor, so Kaplan and company got right to work on the task.

The language specification was created by Kaplan and Ritter to be more than just yet another variant of the popular programming language. Basic09, as it was

Figure 6.2

Microware employees pose for a group photo in the company's lunchroom (circa 1985).

called, was an elegant hybrid of Pascal and BASIC. It would allow for a Pascal-like structured code where line numbers were optional, "gotos" were supported but discouraged, variables were declared with specific types, and callable procedures were the norm. It was unlike any other BASIC language at the time. In addition, the development environment would include a line editor, debugger, and "just in time" interpreter that could emit compact "intermediate code" and be executed by a run-time interpreter. Years later, Java would employ the same model to deploy applications in a platform agnostic manner.

Ambitious as the specification was, the work to implement it was even more so. The daunting task of actually building the language interpreter fell to Kaplan and newly hired employees Larry Crane and Robert Doggett. Working around the clock, the team built the editor, debugger, interpreter, and intermediate code generator piece by piece. According to Kaplan, it was an immense amount of work, and at times, draining. "We almost killed ourselves getting it done," he recalled.

As the work on Basic09 wound down, the team of engineering wizards at Microware realized that something was missing. There was a need for some type of underlying operating system to support Basic09's requirement for a file system and interaction with input/output devices. Having used UNIX on a PDP-11, Kaplan saw the beauty and rationality of the operating system, but realized that "it had some problems." The difficulty in extending the operating system, which was a single blob of code, bothered him, as did the I/O model that he saw as also needing improvement. Kaplan began working on an operating system to address his concerns.

What started out life as a by-product to support the young Basic09, would morph into a new operating system for the 6809 microprocessor and carry on the final digit in its name: OS-9. Kaplan reworked the deal with Motorola, who owned Basic09, since it was a "work for hire" project. Microware made an arrangement to co-own and distribute Basic09 in exchange for OS-9 licenses to Motorola.

The OS-9 operating system soon became the flagship product of the growing business, and requests came in to port the operating system to up and coming 6809 systems (Figure 6.3). On top of that, Microware began to garner an impressive list of language offerings. With this growth came the need for more people, so Microware began to place ads in the local newspaper, *The Des Moines Register.*

One such ad requesting "software wizards" caught the attention of two ambitious young programmers: Mark Hawkins and Kim Kempf. Both Hawkins and Kempf applied for and were offered positions at the growing company, joining the ranks of its earliest employees.

Among Kempf's first tasks: get a C compiler system running under OS-9. Starting with a K&R C compiler written by James McCosh, Kempf and Larry Crane began the task of adapting the compiler to both run under OS-9 and emit executable modules to run under the operating system. A new relocatable macro assembler and linker were also written by the pair.

One of Mark Hawkins's first assignments was to work with Robert Doggett, delving into OS-9's Random Block File manager, or RBF, a module of code that was responsible for performing file and directory management on floppy and

Figure 6.3

OS-9 running on a Fujitsu FM-7, a Japanese home computer series first released in 1982. (Courtesy of Frank Hogg.)

hard drives. This work was in support of the internal user state index sequential file system for Microware's COBOL programming language product. Over time, Microware invested in additional programming languages for OS-9, creating compilers and interpreters for both Pascal and Fortran.

Meanwhile, the folks at Tandy were shopping around for an operating system.

By 1983, the Color Computer was selling well as both a game machine and as a centerpiece for more business-like computing tasks, something that initially threw off the brass at the towers. People were taking the Color Computer out of its entertainment shell and using it for serious tasks like payroll, word processing, and retail sales management, as Tandy's Dr. John Patterson would recall years later. "One of the most surprising things for us was just how much businesses were using those. We just didn't expect that. We really didn't. We had in that time frame, 68000-based UNIX systems and our TRS-80 line. We had some powerful systems... and a lot of outstanding software support. We never expected people to be running their business and accounting on Color Computers. I remember walking into a store and there was a Color Computer sitting on the counter running their business, inventory, and everything. Wow! That shocked me."

Although Radio Shack had Color Computer business software offerings like *Spectaculator* (1981) for spreadsheets and *Color Scripsit* (1982) for word processing, both were somewhat limited cartridge-based applications. There were also disk-based, third-party applications like *Telewriter 64* (1984) from Cognitec and *VIP Writer* (1983) from Softlaw, which brought additional functionality and features over and above what Tandy offered. Yet all of these applications ran under Tandy's limited Disk BASIC environment.

What serious-minded business applications needed was an operating system to manage memory and graphics with a software apparatus that allowed easy expansion and access to peripherals and common software interfaces. Tandy was eager to find that missing piece of the CoCo puzzle. Some in Tandy advocated for the TRS-80's TRS-DOS to be ported to the 6809. Others argued for FLEX, a single-tasking operating system available from Technical Systems Consultants (TSC) out of West Lafayette, Indiana. Ultimately, the choice would be OS-9. Yet, how it came to be the choice operating system for the Color Computer remains in debate and is remembered differently today by those who were in the know at the time.

Whereas Tandy's Mark Siegel recalls enlisting a friend, Mel Norrell, to put him in contact with Ken Kaplan about bringing OS-9 to the Color Computer, Microware's head of sales, Andy Ball had a different recollection. According to Ball, Tandy used a ruse to get Microware to agree to let it evaluate OS-9. As Ball recalls, he was contacted by a professor at Texas A&M who was interested in evaluating OS-9 for a nonspecific 6809 computer. Unbeknownst to Ball, the professor was actually working for Tandy, evaluating both OS-9 and the other popular 6809 operating system, FLEX. The professor's ultimate recommendation to Tandy was OS-9.

FHL's Frank Hogg had already been supporting OS-9 on other 6809 platforms, including the famed GIMIX professional computer systems as well as selling OS-9 software such as *Sculptor,* a powerful database, and *Dynacalc,* a word processor similar to MicroPro's *WordStar,* which was the popular choice on CP/M and PC DOS platforms into the mid-1980s. Frank's nephew, Rich Hogg, began the work of getting OS-9 to run on the Color Computer, otherwise known as "porting" the operating system (Figure 6.4).

Despite the existence of FHL's port, Tandy negotiated directly with Microware to license and sell OS-9 Level One, Basic09, Pascal09, and the Microware C Compiler. Signing with the Texas technology titan to sell Microware's software in its huge number of Radio Shack stores all over the country was a big win for the small Midwestern software company. Kaplan rallied his team together to bring full OS-9 support to the Color Computer. Mark Hawkins would lead the kernel adaptation effort, while other engineers chipped in on bringing up other aspects of the operating system.

Even Microware's president Ken Kaplan became personally involved in the coding aspect of the Color Computer port of OS-9, contributing the floppy disk driver, which ferreted data to and from Tandy's 5.25″ disk system. Kaplan recalls wrestling with Tandy's use of the nonmaskable interrupt in the floppy controller's design and how it hampered real-time performance. The engineer-turned-president did his best to try to minimize the impact on the system to optimize performance while a disk operation was going on, but performance still suffered at times when disk operations were taking place. The most obvious symptom was the lack of true "type-ahead," resulting in missing keys when typing during disk input/output. "It's one area where Microware could have had an impact had we gotten in earlier on the design of the CoCo," Kaplan recalls.

With the port nearing completion in the fall of 1983, Microware's OS-9 was unveiled in Radio Shack's 1984 catalog for $69.95. Soon after, the operating system and its thick bundle of manuals made its appearance at Radio Shack stores.

Figure 6.4

FHL's memorable full-page advertisement selling OS-9 and FLEX for the Color Computer. (Courtesy of Frank Hogg.)

Kaplan still remembers the excitement the first time he saw his product on a shelf at a Radio Shack store. It was truly a moment to remember.

From that point on, OS-9 would usurp all other 6809-based operating systems to become the de facto standard for the CoCo and similar platforms. Usage of

TSC's FLEX would eventually taper off, giving OS-9 a commanding lead in the Color Computer community (Figure 6.5). As time went by, Tandy demanded that more and more of its CoCo disk-based software actually run under OS-9. That insistence, along with the powerful multiuser, multitasking features of the operating system, would carry the Color Computer through some uncertain times ahead.

In the meantime, it was full steam ahead at Microware. One of the perks that the company offered at their headquarters was free lunch for their employees. It's something that Dennis Tanner of Tandy's education department remembers well when he and his boss, Vice President of Education Bill Gattis, visited the software company.

Tandy's expansion into the education market included a push for Color Computers, and no one at Tandy knew more about computers in education than Dennis Tanner. An elementary school teacher in Kansas, Tanner purchased a TRS-80 Model I not long after its introduction and began writing programs for his school. Soon thereafter, he quit his job as an educator to join Tandy as an educational programmer and moved to Fort Worth in June of 1980.

Throughout his 13-year career at Tandy, Tanner went from creating educational titles to managing Tandy's half-dozen programmers and educational content developers. According to Tanner, Tandy invested heavily in education, and the Color Computer was one of its centerpiece computers for schools. It was inexpensive, could be expanded as needed using cassette- or disk-based storage, and had color graphics. Educational software like *Vocabulary Tutor 1*, *Crosswords*, *Bingo Math*, and *Fun With Reading* focused on the famous three R's.

Science-based software was also represented on the Color Computer with *Atom*, a clever game that taught students about the basics of molecules, and *Color Computer Learning Lab*, an eight-cassette tape-based learning series. Tandy's Bill Gattis even went as far as finding and convincing American astronomer and discoverer of the planet Pluto, Clyde Tombaugh, to participate in the development of a program for the CoCo that documented the then ninth planet in the solar system, *Discovering Pluto with Dr. Tombaugh*.

In addition to overseeing Tandy's internal educational titles, Tanner also worked with companies such as Norman, Oklahoma-based Dorsett Educational Systems. Together they marketed programs based on the Talk Tutor platform, a system developed by Jamie Alexander for the Color Computer. *Talk/Tutor* used its cassette tape as both a program and audio source, intermingling both and allowing interactivity. This technique became the basis for several Color Computer programs spawned from a collaboration between Tandy and Walt Disney. Disney titles introduced in the 1984 Radio Shack catalog and beyond included *Telling Time with Donald*, *Problem Solving with Scrooge McDuck*, *Mickey's World of Writing*, *Goofy Covers Government*, *Donald Duck's Playground*, *Mickey's Alpine Adventure*, *Math Adventures with Mickey*, *Space Probe: Reading*, and *Space Probe: Math* (Figure 6.6).

In the same catalog, a number of educational titles from the creators of the famous children's television show *Sesame Street*, Children's Television Workshop (CTW), appeared. Children's Computer Workshop (CCW) designed titles for Apple, Atari,

Figure 6.5

The announcement for Microware's 3rd Annual OS-9 Users Seminar, held August 1984 in Des Moines, Iowa.

Figure 6.6

A few of the Disney CoCo titles are shown. Some titles were in partnership with Sierra, as well as a number of CTW/CCW software packages. Several of these titles received the Tandy Home Education Systems branding.

and Commodore, and decided that they would adapt many of those programs for the CoCo. One program, *Taxi*, allowed kids to simulate driving around a city. If the child was speeding, a cop would come out from behind and stop the vehicle. Other titles released included *Star Trap, Grover's Number Rover, Peanut Butter Panic, Ernie's Magic Shapes, Big Bird's Special Delivery, Cookie Monster's Letter Crunch,* and *Creative Exploration Series: Grobot, Timebound, Flip Side.*

One of Tanner's new hires was Keith Moore. Moore was brought on to do both BASIC and assembly language programming of educational titles for Tandy. One impressive feat of programming that Moore was particularly proud of was a nationwide "mock election": in high schools corresponding to the 1984 presidential race. During the evening of the election, schools across the United States coordinated to send student votes to a state computer, which would then forward that data to a national computer. Tandy sponsored this program and the Color Computer was used to aggregate and display that data. Moore recalls that the results of the mock nationwide school election had very similar results to the real election: incumbent Republican Ronald Reagan defeated Democrat Walter Mondale 58.8% to 40.6%.

In addition to Tandy's focus on education, there were a plethora of third-party companies who catered to the CoCo in that arena. Educational software company TCE offered a wide variety of packages focusing on early learning, math, language arts, and even word processors, data managers, and spreadsheets for children:

Child Writer, Child Filer, and *Child Calc. THE RAINBOW*'s regular contributor and teacher Steve Blyn also advertised preschool and K–12 educational software through his company Computer Island out of Staten Island, New York.

So important was the mission of education to the CoCo community that Falsoft dedicated its annual September issue of *THE RAINBOW* to education, focusing on articles and programs to benefit K–12 and even college-level students. Individuals in the wider CoCo community were known for their focus on educational efforts, including the aforementioned Steve Blyn, Fred B. Scerbo, and Michael Plog, Ph.D.

The continued success of the Color Computer 2 as a machine for education, gaming, and serious endeavors continued during the late 1984 holiday shopping season. Tandy's marketing head Ed Juge was quoted as saying, "Each year, Christmas sales of the Color Computer break the previous year's record." Such a trajectory demonstrated to Tandy that the CoCo market was still vibrant. However, it was clear that the limited graphics and text resolution of the CoCo 2 needed to be addressed. Since its inception in 1980, the Color Computer had lived with a restricted 32 × 16 character screen and reverse video characters used to represent lowercase. Although third-party solutions attempted to raise the bar with kits that added true lowercase characters and other video enhancements, not having at least a 40 × 24 text screen built-in limited the CoCo's potential.

Graphics were also a limiter. The four-color pseudo feature that depended on the nuances of the NTSC television signal could take the CoCo only so far with its 256 × 192 resolution. Compared to other low-priced home computer systems on the market at the time, including the Commodore 64 and family of 8-bit Atari computers, the CoCo 2 was at risk of losing out on sales due to this limitation alone, which was obvious to just about anyone after even just a casual comparative glance.

Something would have to be done, and soon.

7

Three's Company

With the Deluxe Color Computer scrapped and the CoCo 2's capabilities beginning to stale in comparison to its competitors, Tandy's Mark Siegel began anew to sell the idea of a worthy successor to Radio Shack's Color Computer buyer Barry Thompson. As the dynamic duo hashed out the details, their vision began to take shape into something that would significantly eclipse the current limitations of the Color Computer and quench the thirst of the enthusiasts whose growing choir of voices demanded a more powerful machine.

From Tandy's point of view, any successor to the Color Computer 2 would have to, as a matter of course, maintain compatibility with the previous machines. It was important to preserve the company's investment in the hundreds of software titles that had been written and developed over the years. All software that ran on the CoCo 1 and CoCo 2 would need to run on a new and improved Color Computer, irrespective of any changes to the graphics or sound capabilities. This would naturally mean that certain hardware features would automatically carry over, such as the 40 pin expansion connector on the side, as well as the joystick and cassette ports on the back, and the infamous pseudocolor mode. It also meant that the venerable 6809 microprocessor would remain front and center as the CPU workhorse.

Another requirement was that the new machine would have to run OS-9, Tandy's chosen operating system for the Color Computer. The investment made in OS-9-based software available through Radio Shack was already solid and still growing, so Tandy's commitment to the multitasking, multiuser operating system from Microware was unwavering. The new machine simply had to run OS-9.

In consideration for OS-9 running optimally, it was determined that the CoCo's text resolution would have to be increased. Since compatibility was important, the 32 × 16 screen made popular by Motorola's 6847 VDG chip would stay, but additional 40 × 24 and 80 × 24 text modes would also be available.

Increased memory was yet another hard requirement. Even though as an 8-bit microprocessor the Motorola 6809 could only address 64K of memory at a time, it could be coaxed into using much more through clever banking schemes. Microware's OS-9 Level Two product, which ran on other 6809-based platforms such as the GIMIX, took advantage of larger memory by working with memory management hardware that fractured large RAM up into a number of smaller, manageable blocks. A typical OS-9 Level Two system might have 512K of RAM with 2K blocks, yielding 256 individual pages that could be assembled to provide up to eight unique 64K address spaces for the 6809.

Generally, the smaller the block size, the more efficient the use of memory, but it came at a cost: denser, more complex logic, which tended to raise the price for the part. Tandy engineer John Prickett discussed the proposed memory specifications with Microware: 128K of RAM would be standard on the CoCo 3, expandable up to 512K. Microware suggested that the CoCo 3 have 2K or 4K pages, but relented when Prickett recommended and got 8K pages in order to keep costs down.

In addition to improving the available memory and text screens, one of the areas most obviously ripe for enhancement in the new machine was its graphics. Having witnessed the home computer and game console market shifting toward increasingly sophisticated graphics and sound, Tandy knew that this had to be addressed. Earlier, Motorola had approached Tandy with the idea for a new graphics chipset. Originally designed as a solution for the 68000-based Tandy 6000, the design was intended to go into direct head-to-head competition with graphics offerings for PC compatibles of the era.

This new graphics and memory management chipset, known as the Raster Management System, or RMS, was composed of two separate integrated circuits: the MC68486RMI (Raster Memory Interface) and the MC68487RMC (Raster Memory Controller). Together, they would provide generous 64 on-screen colors from a palette of 4,096, closely mirroring the vaunted capabilities of Commodore's high-end 16-/32-bit Amiga 1000 computer. Although targeted for the 68000 series, Motorola insisted that the RMS could also work with the 6809. So serious was the effort that Motorola put into the RMS that it assigned some 30 people to the project.

With the path set, the wheels were in motion to begin the work of building the new and improved Color Computer.

For any complex circuit design undertaken at the time, a multitude of discrete components were initially used to develop a prototype for the chip it was to become. With the RMS, it was no different. The early RMS prototype was composed of wire-wrapped ICs planted on multiple boards that filled a sizable box, and drew significantly more electrical current than the final chipset was expected to use (Figure 7.1).

Anxious to gauge progress, Siegel visited the RMS's design team at Motorola's location in Arizona several times. As work on the design progressed, he came to the conclusion that the RMS simply would not fit the vision that he had for the CoCo 3. As Siegel explained, Motorola's design used a memory register, which in effect acted as a sliding pointer to partition RAM into two banks: video RAM and nonvideo RAM. Tandy engineers had envisioned multiple "blocks" of equal-sized memory that could be mapped in and out of the 6809's 64K address space, a model that closely fit the OS-9 operating system's way of managing memory.

Concern about the memory management issues and its divergence from OS-9's specific requirements came to a head at a meeting among the engineers and several PhDs from Tandy's research and development department. Two camps quickly formed in the room: those who saw the RMS as a dead end, and those who wanted to continue to pursue integration of the chipset.

Motorola asked Tandy if they could just live with the design as it stood, but the engineering team held firm to its mantra that any graphics and memory solution had to be compatible with OS-9. Motorola finally admitted to the intractability of OS-9's memory requirements vis-à-vis the RMS, resigned to the fact that

Figure 7.1

The RMS prototype cage. Had the project not been canceled, all of these chips would have been scaled down to just two integrated circuits. (Image courtesy of Ellis Easley.)

OS-9 Level Two, as designed, could not access memory in such a way. "There are problems that have no resolution," recalled Siegel.

Tandy's engineering manager Dale Chatham also remembered the contemplation of using the RMS in the design and believes that it was dismissed almost as soon as it was suggested. "Most likely due to cost, as Tandy was very much cost driven in the product development cycle," Chatham recalled, "I think it was dismissed at almost the same time as the idea of possibly using the 68000 for the CPU." This alluded to another design consideration that was dropped early on: using Motorola's 68000 processor instead of the 6809. "The 68000 was definitely dropped due to cost. Also, I do remember being disappointed that Motorola decided not to enhance the 6809."

Cost was indeed a factor, as John Prickett would confirm years later. At around $20 for the two-chip set, it would take a huge bite out of the budget set for the machine. But aside from price, Prickett also saw a significant flaw in the design: the RMS could only run at .89 MHz. Why not push the 6809 to 2 MHz and double its processing power? "I was adamant about doubling the processor speed," Prickett explains.

With the issues mounting, Prickett confidently told his colleagues at Tandy that he could "make a graphics chip, and a lot cheaper too."

Prickett knew that building a graphics processor would be a daunting task, as it meant going the route of creating an application-specific integrated circuit, or ASIC, something that took significant time and engineering resources to do properly. More important, it was not something that he had done up to that time. Prickett saw the opportunity to design an ASIC as challenging and at the same time exciting.

A preliminary plan was put together and provided to Tandy management, including Bernie Appel and John Roach, who gave their blessing.

Chatham and Prickett started working on a new set of specifications with input from Siegel. One of the first things to go was the 4,096 color palette. As Prickett would later recount, "4,096 colors was cool, but would have taken a 12-bit DAC [digital-to-analog converter], which would be expensive. We could do 64 colors with a 6-bit DAC."

In addition to being able to display up to 16 colors on-screen at one time from its total palette, the proposed chip would also yield a maximum resolution of 640 × 192, with additional resolutions of 320 × 192 and 160 × 192. There would also be support for 40 × 24 and 80 × 24 hardware text screens that supported underline, blinking, and color attributes for each character. Smooth graphic scrolling and foreground/background bitmaps were also key to the design as was a keyboard interrupt and a 12-bit programmable timer interrupt.

Although the video specifications were off in some ways from the original target, they would still enable the CoCo 3 to match up favorably with the most common resolution and color settings of the era's newest and most popular high-end computers. These competitors included the Atari ST, which typically displayed 16 out of 512 colors; the Commodore Amiga, which typically displayed 32 out of 4,096 colors; IBM PCs and Compatibles equipped with EGA graphics cards, which typically displayed 16 out of 64 colors; and Tandy's own 1000 series computers, which typically displayed all 16 of its available colors.

With specifications firmed up, the small team began the arduous task of creating what Prickett would call the "CoCo 3 chip." Chatham performed the initial product specification, then handed his work off to Prickett, who took care of all of the detailed specification work as well as the detailed design work on the chip. To start, the CoCo 3 chip would be prototyped from discrete components.

One question vexed Prickett: "How do we get from the digital realm inside of the chip to the analog color that has to drive the monitor?" Drawing from his experience as a TV repairman, he began by designing an NTSC generator inside of the chip as well as the composite video machinery. This required Prickett to duplicate the horizontal and vertical sync signals all in the chip as well. For CoCo 2 compatibility, the functionality of Motorola's 6847 VDG would also have to be incorporated into the new circuit.

As the design progressed, it became obvious that there was a problem with the color space between the analog RGB output and the composite/NTSC output. "It was physically impossible to map the color set the same between the two," Prickett recalled. Other compromises had to be made in the construction of the available 64-color palette. In one of the color modes, Prickett revealed that the color white was duplicated, providing only 63 unique colors.

As the design for the graphics hardware continued to progress, attention turned to improving the sound of the CoCo 3. Up to that point, the audio capability of the CoCo consisted of utilizing the 6-bit DAC to generate sound and tones. Except for the Apple II series and unexpanded IBM PCs and Compatibles, no other major computer platforms active at the time featured such limited sound capabilities, and certainly none of the newer systems that the CoCo 3 would be competing against. Cartridge-based sound hardware such as the Speech/Sound Pak or the Orchestra-90 CC were capable solutions, but they were not built in, limiting software support. CoCo users yearned for advanced music and sound capabilities built right into their Color Computers.

The popular General Instrument AY-3-8910 Programmable Sound Generator, whose variations were used in everything from the Mattel Intellivision game console to the Mockingboard sound card for the Apple II to the Atari ST and even Tandy's own Speech/Sound Pak—as well as the never-released Deluxe Color Computer—was eyed as a potential chip. With 3-voice sound and the ability to produce 4,096 different pitches, the General Instrument chip would go a long way toward making the CoCo 3 more appealing for games and sound applications.

Another component in dire need of addressing was the built-in communications port. Every Color Computer system sported the 4-pin DIN connector that allowed interfacing to serial devices such as modems and printers. Limiting the efficacy of the port was the fact that the assemblage of bits coming down the wire had to be done in software, a time-sensitive task that took the CPU away from other important tasks. Having a hardware-based serial solution would absolve the 6809 from having to work so hard just to send and receive data, and would allow connectivity to much faster devices, far beyond the limiting 1200 baud that the bitbanger port could do. Much like the sound options for the CoCo, while there was a Deluxe RS-232 Program Pak that provided an improved serial port, it too was an optional add-on cartridge.

After exploring the available options to enhance the sound and serial communications, it soon became clear to Siegel and the engineering team that the cost constraints of the product's budget would not allow either feature, let alone both. "I decided that Tandy could not live with either," Siegel recalled. "The only drawback [to the existing sound] was the interleaving of the access to the DAC to maintain sound stream continuity." The CoCo 3 would carry on both the legacy of 6-bit sound, and the software-driven serial port, taking some of the luster off an otherwise solid series update.

One area that Prickett saw as a deficiency was the processor speed. Even a powerful processor like the 6809 needed a boost, and at a paltry .089 MHz there was not enough horsepower to run OS-9 the way it should. Prickett lobbied for a faster version of the processor. Siegel went for broke, asking the team at Motorola about the possibility of ramping up the 6809 to run at a whopping 4 MHz. The response from Motorola was terse: "We cannot violate the law of physics." Eventually, Tandy settled for a doubling of the clock speed to a more realistic 1.78 MHz and incorporated the 68B09E, Motorola's 2 MHz version of the CPU.

With the design coming together, it was time for prototypes. Prickett sent off his design to CUPLEX, a printed circuit board manufacturer in Garland, Texas. Some time later, the boards were there to greet Prickett when he arrived to work one Friday morning.

The first task that Prickett faced was to rewire the board to accommodate a change of RAM that was mandated by Tandy. Although slated to leave for a vacation the next day, the dedicated designer was eager to see his creation come to life. Working 25 hours straight, he finally saw the colored cursor appear on the screen on Saturday, delaying his vacation by just a day.

With the new graphics, text, and memory capabilities specified for the CoCo 3, it was clear that BASIC would need to be extended to take advantage of these features. Naturally, Tandy management approached Microsoft to see what could be done. To Siegel's surprise, not only did Microsoft show no interest in providing a new version of BASIC for the CoCo 3, it also presented Tandy with a seemingly impossible scenario. According to Siegel, Microsoft stated, "You [Tandy] cannot create another incarnation of the CoCo unless it has Microsoft BASIC built into it. And we [Microsoft] aren't going to do the work." Although Microsoft did not give a specific reason for its seeming contradiction, Siegel contends that Microsoft took the stance to force Tandy to concentrate on its Tandy 1000 PC compatible line, which was much more lucrative for the Redmond, Washington-based company.

Undeterred, Siegel began to think of a way for the CoCo 3 to have its cake and eat it too. While it was strictly forbidden to provide a modified BASIC ROM, someone—or some company besides Microsoft—could develop a ROM that would patch BASIC on startup. This would technically not modify the ROM but would copy the contents of the ROM into RAM and then patch in the enhancements. It was a fine line to walk, so Siegel immediately set to work to see just how it could be tackled. The ultimate solution would lay with another engineer, a Microware employee named Tim Harris.

Harris got his Color Computer while attending Iowa State University, where he was introduced to OS-9 on his Color Computer and eventually switched majors to computer science. While working on his degree, Harris wrote an address book application in Basic09 that found its way into *Dr. Dobb's Journal*, which got the attention of OS-9 aficionado and author Dale Puckett. Puckett asked Harris to contribute sample code to his upcoming book, *The Complete Rainbow Guide to OS-9*, which was coauthored with Peter Dibble.

With such impressive credentials, Harris applied for and received an employment offer from Microware. The company wanted to hire Harris before he finished college, but Harris instead insisted on completing his education. He graduated six months later, in December of 1984. Meanwhile, another company came calling for Harris: Microsoft. After some thought, Harris shrugged off Microsoft's offer and instead became Microware's 14th employee on January 3, 1985.

Harris's first project as a new hire would come from one of Microware's founding engineers, Larry Crane, to write a Fortran compiler for Microware's customer, the Iowa Department of Transportation. Having written a Pascal compiler for his undergraduate work, Harris was a natural for the project. As he would later recall, though successful, the design was arduous, "I had a diagram of the project on a whiteboard that took me weeks to assemble; had it been erased, the whole project would have been dead."

After moving on from the Fortran compiler, Harris got his first taste of the work that inspired him to go with Microware in the first place—developing for his beloved Color Computer. He was tasked with writing the Speech/Sound Pak and serial drivers for the soon-to-be released OS-9 Level One Version 2, then moved on to writing the firmware for a networking board that plugged into the CoCo's expansion port. Designed for schools, the networking board connected via twisted pair wire to other CoCos in a loop configuration, with one machine acting as the master. "It was a pain in the butt to get it to work," Harris recalled, which was no surprise, considering that a single malfunctioning computer in the ring could create problems for the whole network.

As Harris dutifully labored to bring new features to the Color Computer's signature operating system, he began to hear rumblings down the halls in Des Moines. People at Microware were talking about a new machine that Tandy was working on, something that would eclipse the Color Computer 2 in both speed and graphical capabilities. Having read Tracy Kidder's landmark 1981 book *The Soul of a New Machine*, which chronicled the experiences of an engineering team as it raced to design a next-generation computer, Harris quickly realized the significance of working on a project such as this. He would have that opportunity sooner than he would think.

Rumor and speculation gave way to reality as winter came to the heartland in November 1985. That month, Microware president Ken Kaplan, along with Harris and a fellow engineer and Iowa State graduate Mark Hawkins, flew down to Fort Worth to meet with Tandy's engineering and marketing groups. There they were summoned to discuss grand plans for a series of secret projects. In addition to having Microware bring its OS-9 Level Two operating system to the CoCo 3, Siegel had specifications for a new graphical interface application dubbed "Multi-Vue," so the idea of asking Microware to write the extensions to Microsoft's Color BASIC was a natural one. Kaplan committed Microware's resources to the work ahead, including writing what would become known as Super Extended BASIC. Not only would the work mean more revenue from Tandy, but it would also steal a win from Microware's perceived arch nemesis, Microsoft. With Microware's participation in the creation of the CoCo 3 looking solid, Hawkins and Harris would add a third Microwareoid to the team: Todd Earles. This trio of engineers would play a significant role in ushering in what would be known as the Tandy Color Computer 3.

In December, just one month after visiting the Tandy Towers, the cowboys of Fort Worth returned the favor with a visit of their own. When they arrived at 1900 114th Street in Des Moines, they carried precious cargo consisting of two 24″ × 36″ plywood boards, each containing over one hundred wire-wrapped discrete logic components.

The boards were early prototypes of the yet-to-be-finalized CoCo 3. The enhanced graphics/memory chip, which would eventually be shrunk onto a PLCC packaged chip the size of a postage stamp, was for now represented by dozens of individual chips on the prototype board.

The gauntlet had been thrown: Microware had committed to expand a BASIC language, bring over a new operating system, and design a graphical user interface, all in the space of 18 months. The work was split between Harris, Hawkins, and Earles. The extension to the BASIC ROMs as well as OS-9 graphics drivers would be Harris's domain, while Hawkins would bring up OS-9 Level Two on the new hardware. Earles would focus his efforts on Multi-Vue and other graphical components of the operating system.

As Hawkins worked to bring OS-9 up on the prototype, Siegel mapped out a windowing system that would allow users to fully exploit it. The operating system would support both hardware and graphical (bitmapped) text screens. Using simple escape codes, windows of different sizes could be created, with a different application running in each.

Meanwhile, Harris, mindful of the challenge that Microsoft presented Tandy, got to work looking at how to safely and effectively patch the BASIC ROMs at startup, without actually modifying the ROM itself. Thanks to Spectral Associates and their "Unravelled" series of books, which provided fully disassembled, annotated listings of the complete set of BASIC ROMs for the Color Computer, much of Harris's work was already done. Studying the listings gave him the edge he needed to determine where to patch the ROMs to allow the new set of BASIC commands.

Development was challenging, however. The physical size and the sheer number of chips made it difficult to keep the prototype boards running reliably (Figure 7.2). "We had to use cans of Freon to keep the boards cool," recalled Harris, "otherwise, the screen would start going haywire and eventually the

Figure 7.2

The prototype bare-board CoCo 3 used by Microware to develop the BASIC ROM. (Image courtesy of Allen C. Huffman.)

board would lock up." Even just slightly flexing the board would risk malfunction. In addition, having all of the components exposed greatly increased the chances of permanently damaging the boards. Even with great care being taken, Harris himself inadvertently "fried" one of the prototype boards. It was not until later, in 1986, that nearly complete CoCo 3 motherboards in the familiar white cases arrived in Des Moines, which made development a little less stressful.

As the development of the BASIC enhancements were underway, Harris realized that there would be a lot of wasted space in the 32K ROM that was marshaled to hold both the Color BASIC and Extended BASIC ROMs as well as the Microware patches for the BASIC enhancements. When asked what other features or capabilities could be placed in the additional ROM space, Tandy management replied, "Just fill it with random junk bytes." And that is exactly what the engineers at Microware did, albeit in a clever and unique manner.

Back at the Tandy Towers, nearly complete CoCo 3 prototypes that actually looked like Color Computers, complete with case and keyboard, were being manufactured. It was time to bring in a Dallas company, VLSI Technology, to shrink the design to a size that would fit concisely on the new prototype motherboard.

Jim Bruister was VLSI's design engineer tapped to work with Tandy's John Prickett. The pair immediately hit it off when they learned that each had a penchant for music. Bruister, a tuba player, was fondly called "Tube" by Prickett, who himself was a former music major. Together, they began the work of taking the breadboard and shrinking it onto silicon that would be barely larger than a postage stamp. It was an experience that gave Prickett deep insights into how a large-scale integrated circuit was put together. "I learned a lot about chip design from Jim," Prickett recalls years later.

Dale Chatham also gave credit to Tandy's Tokyo engineering group, who took on the unsolicited task of reverse engineering the SAM chip. "They sent the Fort Worth engineering group a schematic of that logic," recalled Chatham. They were hoping to save money by getting a lower cost version of the chip. However, that never happened. "I am sure that John saw this schematic to at least get a few clues to help with his design of the GIME [Graphics Interrupt Memory Enhancement] chip," speculates Chatham.

As successive iterations of the design were worked and reworked, the chip became more solid and stable. When a semifinal set of ceramic chips came in for the new prototypes, they were indeed barely bigger than a postage stamp—but there was a problem. The fabrication process had apparently left a little too much metal on one layer inside of the chip, adversely affecting video performance. After some consultation with the manufacturer, it was determined that applying a certain amount of voltage between two pins on the chip would cause enough energy to "burn away" the excess material on the layer. Prickett asked Tandy engineer Ellis Easley to lend a hand on this unusual task.

Another issue was found with early prototypes of the GIME. When they arrived, Prickett eagerly plugged them into the carriage socket and turned on the CoCo 3 to discover . . . a black screen. He clicked the power switch to turn off the apparently nonworking CoCo 3, only to view a faint, green screen appear for a brief amount of time before vanishing. After doing this several times, he discovered a flaw. The chip's DACs could only tolerate 3.5 volts. At 5 volts, the DAC was saturated and all of the colors went black. With a description of the problem, Prickett took his findings to VLSI, which found the problem and started a three-week turnaround for a new chip. Fortunately, the replacement chip worked.

Other challenges existed. While the 32 character screen mode was fine, Prickett discovered that in 80 character mode he was half a character off. Eventually that too was fixed and VLSI turned out another revision of the chip.

After many arduous months of design work, fabrication, testing, and rework, the final set of test chips came in. Stamped with the year "1986," the new GIME chip was complete. Soon, many thousands of clones would be manufactured and shipped overseas to be placed into the sockets of expectant motherboards sitting on factory tables in South Korea.

Years later, Chatham would praise Prickett's work on the GIME chip. "Developing an ASIC is a huge amount of work, and John deserves *all* of the credit for the GIME chip. If I had been assigned the task of developing the chip, I am sure it would not have been as successful or have had as many features as John's design. A bigger company like Motorola would have had teams of several engineers designing the chip, and I doubt that it would have been as good as John's design."

In the end, the loss of the RMS chip was a coup for Tandy. A bright engineer stepped up and saved the day by designing a superb chip that would propel Tandy's latest Color Computer to new capabilities. As a bonus, the bean counters at Tandy had further reason to smile: the cost of the GIME came in under $7, nearly a third of the cost of the RMS solution that Motorola had proposed just a year earlier.

Toward the end of the CoCo 3 project, Prickett had gotten wind that Tandy was not interested in doing anything more with the 6809. Not finding any inspiration in working with Intel's increasingly dominant architecture and discouraged with the change of direction, Prickett left his job at Tandy in January 1986 to take on a new position with a startup to design a state-of-the-art pager. His colleague, Ellis Easley, would pick up the slack needed to ready the CoCo 3 for final production, quality assurance, and, eventually, release.

On July 30, 1986, the culmination of almost two years of work by dozens of individuals came to a crescendo when Tandy officially unveiled the Color Computer 3 at the Waldorf-Astoria Hotel in New York City, at a retail price of just $219.95. With its impressive color palette, graphics resolution, and text screens as well as 128K of expandable RAM, the Color Computer 3 appeared to be almost everything that CoCo enthusiasts were hoping for. To top off the new features, the CoCo 3's video circuitry could drive either a standard composite monitor or the new high-resolution Tandy CM-8 analog RGB monitor, which allowed the computer series freedom from the fuzziness of a television display for the first time. Things were indeed looking up at the Waldorf-Astoria.

Among the attendees that day were Tandy's John Roach, Mark Yamagata, Ed Juge, Barry Thompson, and Mark Siegel. THE RAINBOW magazine's Lonnie Falk and Jim Reed were also in the room to welcome the CoCo 3 into the world. Rounding out the list of notables in attendance—and in what could only be labeled as irony—Microsoft's own Bill Gates was on hand as well to witness the announcement.

In September 1986, Radio Shack's 1987 RSC-17 catalog began appearing in customers' mailboxes as well as in Radio Shack stores and dealers across the country. For Tandy computer enthusiasts, including Color Computer users, this annual event was particularly important, because it heralded new hardware, software, and accessories.

Page 37 of RSC-17 was completely devoted to the new Tandy Color Computer 3. Posed immaculately with the new CM-8 RGB monitor and sheets of multicolored paper laid out on the table underneath, the CoCo 3 looked particularly inviting. The generous 128K of RAM, higher density 40 × 24 and 80 × 24 text modes as well as the increased graphics resolution and 64 color palette were all highlighted in the accompanying text, making for a tantalizing read for enthusiasts.

As the Color Computer 3 began to appear in Radio Shack's catalog and flyer mailings that fall, CoCo fans young and old called or made their way to their local Radio Shack stores to preorder. The September 1986 issue of THE RAINBOW magazine featured the CoCo 3 prominently on its cover, bursting out of the page as though it was blasting into the stratosphere. Several articles from well-known game designers, including Greg Zumwalt and Dale Lear, graced the pages of that issue, building excitement even more.

The first RAINBOWfest since the introduction of the CoCo 3 took place October 17–19 at the Hyatt Regency in New Brunswick, New Jersey. The pent-up energy of the event was specifically geared toward the new machine and its potential. According to "The RAINBOWfest Reporter" column in the February 1987 issue of *THE RAINBOW* magazine, which highlighted the event, "People appeared in the exhibit hall in waves. Just as you started to move down an aisle, it would crowd up and you would be blocked." The CoCo 3 was billed as RAINBOWfest's "Top Dog."

That same RAINBOWfest featured a CoCo 3 "Round Table" event, which assembled a panel to discuss the new CoCo 3 on Saturday evening. Panel participants included Tandy's Barry Thompson and Mark Siegel, along with Steve Bjork, Dale Lear, and Lonnie Falk, with each taking questions from eager audience members. Everyone was anxious to learn more about the new CoCo 3 and its forthcoming operating system, OS-9 Level Two.

Even though CoCo 3s were being sold at Radio Shack's booth during the event, the availability of the product up to that point had been limited due to a miscalculation by U.S. Customs agents. As Thompson explained, Radio Shack had worked out an agreement whereby the product would ship directly to their various regional warehouses. U.S. Customs would clear the shipments there at the location. Despite trying to stagger the arrival of batches of product at each warehouse so that Customs agents would have time to clear warehouses in a consecutive fashion, the CoCo 3s arrived at all warehouses at the same time. This delayed the clearing process, and, as a result, stymied the availability of the computer at individual stores.

The lead up into the 1986 holiday season appeared to bode well for Tandy. Radio Shack stores were beginning to receive the new CoCo 3 in larger quantities, allowing them to be set up in the store and showcasing the matching CM-8 RGB monitor and floppy disk drive system. With the product showing up at stores with regularity, an increasing number of eager fans got to open their own white boxes with red lettering to find a shiny new CoCo 3 with an RF switch box and cable, BASIC Manual, and a complimentary copy of *THE RAINBOW*. Falsoft's Lonnie Falk and Jim Reed had made a deal with Radio Shack to include copies of the magazine in an effort to both boost readership and to give new Color Computer owners immediate exposure to the fascinating and well established CoCo community.

One of the few outside of Tandy to receive early production runs of the Color Computer 3 was prolific game and software house Spectral Associates, of Tacoma, Washington. Tandy asked the company to create a demonstration program that would highlight the features of the new computer. David Figge and John Gabbard of Spectral Associates went to work, creating a very impressive, multiscene story of the building of the CoCo 3. The demo was written as a volunteer effort by the company and featured colorful vignettes, including a rainbow with cycling colors hovering over a mountain scene, and a crane constructing an entire CoCo 3 system, including Multi-Pak, disk drive, and printer. Music and sound accompanied the demo, combining with the visuals to admirably showcase the new computer's capabilities. Figge and Gabbard would later write another CoCo 3 demo for Radio Shack with a Christmas theme (Figure 7.3).

Back in Fort Worth, however, there was concern. Something was wrong with the CoCo 3.

Testing of the production GIME had shown an anomaly in the behavior of the keyboard interrupt. Instead of receiving a single interrupt when a key was pressed, multiple interrupts would start to fire, resulting in repeating keys that would go on ad infinitum. There were also reports in the field that the horizontal graphics scrolling feature was not working properly.

As fate would have it, John Prickett would return to Tandy just in time to diagnose and solve the problem. His leave from the company back in January 1986 to take a position with the small pager startup had not gone so well. The company went bust, and Prickett came back to his former employer 50 weeks later, in the middle of January 1987. "I re-interviewed with Bill Wilson," Pricket recalled. "I

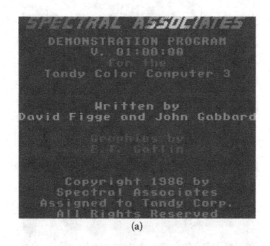

(a)

(b)

Figure 7.3

Screenshots from the two CoCo 3 demos made by Spectral Associates for Radio Shack stores. (Images courtesy of L. Curtis Boyle.) (continued)

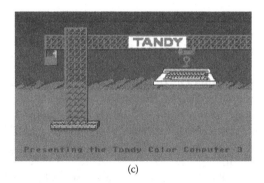

(c)

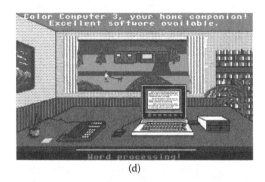

(d)

(e)

Figure 7.3 (continued)

Screenshots from the two CoCo 3 demos made by Spectral Associates for Radio Shack stores. (Images courtesy of L. Curtis Boyle.)

believe I got my old job back at Tandy, in part, because of the problems that they were having with the [GIME] chip."

With Prickett back in the saddle at Tandy System Design, he began to research the problems. As he began to go down the rabbit hole with VLSI Technology, he uncovered an internal race condition that was eventually traced back to the fabrication process. VLSI had created the design and mask, but the production chip was built by a third-party manufacturer. Once the problem was addressed, a new batch of GIME chips, stamped with the year "1987," came off the production lines (Figure 7.4). The existing CoCo 3s out in the field to that point, however, were not recalled. To this day, there are still CoCo 3s to be found with the earlier, buggy 1986 GIME chip.

The problem GIME was eliminated, and Prickett again found himself facing a future without working on his beloved Color Computer. However, he now held knowledge of ASIC design and garnered valuable experience from the CoCo 3 project that would prove to be critical to Tandy. Despite his less-than-fond feelings for Intel's microprocessors, he went on to work on Tandy's Model 1000 line of computers, where he designed the system integration controller chip for the Tandy 1000SX and other systems in the series.

The father of the CoCo 3 had moved on, but there were still a few surprises left to be found.

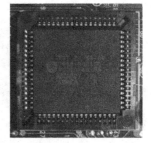

Figure 7.4

From L–R: The progression of the CoCo 3's GIME chip: a prototype of GIME, the 1986 GIME, and the final 1987 GIME.

8

Third Time's Charm

Jim Reed, managing editor of *THE RAINBOW* magazine, penned a curious column in the December 1986 issue, noting the existence of an "Easter Egg"* hidden in the innards of the newly released CoCo 3. "Free the CoCo 3," as Reed put it, would be a contest for readers of the magazine to see how they could call up the esoteric picture of what would become known as "the Three Mugateers" with the mysterious names "M. Hawkins, T. Harris, and T. Earles" (Figure 8.1). Unbeknownst to Reed, he had publicly opened a Pandora's box that had been ostensibly shut months earlier by Tandy officials.

Not long before the announcement of the CoCo 3 in the summer of 1986, 50,000 units sat idly by at a Tandy factory in South Korea, ready to ship to Radio Shack's regional warehouses in the United States. A few units made it in early to Fort Worth, Texas, for testing, including to Mark Siegel, who had led the development effort. Shortly thereafter, Siegel received a telephone call from Mark Hawkins, Microware's engineer responsible for the OS-9 Level Two port. "Hold down the CTRL and ALT keys, then press the RESET button," Hawkins told

---

* Easter Eggs are hidden messages, secrets, or extra features in a videogame or other computer software. Though first popularized by Warren Robinett's *Adventure* (1979) on the Atari 2600 VCS, Easter Eggs have been a part of the computer history's earliest days.

Figure 8.1

The infamous "Three Mugateers" graphic embedded in the ROM of the CoCo 3. The original individual photos were combined to create the infamous Easter Egg. (Compare to Figures 8.2 to 8.4.)

Siegel, who then dutifully followed the directions. What appeared on Siegel's monitor at first took him aback, then set him into a silent, seething rage.

"I saw the picture, and I was really, really pissed," recalled Siegel as he recounted the story years later. "You have no idea how mad I was." Yet, despite his anger, Siegel refrained from admonishing Hawkins on the phone. Instead, there was a period of uncomfortable silence. "I didn't say anything," he recalled.

Finally, Siegel gathered his thoughts and told Hawkins simply, "All right, I'll get back to you," then abruptly hung up the phone.

On his own, Siegel did not really have a problem with the Easter Egg, but at the same time, he knew that he would have to explain the shenanigan to upper management, and that is what he dreaded most. With so much product already assembled and deadlines looming, it was too costly a proposition to replace the offending ROM chips. Resigned to being forced to share the news, Siegel called his boss, Mike Grubbs, and Radio Shack's Color Computer buyer Barry Thompson into his office, where he demonstrated the "feature" that Hawkins had told him about a short while earlier. Grubbs and Thompson were stunned.

Despite the fact that 50,000 CoCo 3s were sitting in a South Korean factory, ready to be boxed and sent to the states, the trio began to consider the possibility of canceling the entire project. "We were this close to killing the product," Siegel remarked, holding his forefinger and thumb at an uncomfortably close position to magnify the effect. "This close."

After some ensuing discussion, Siegel, Grubbs, and Thompson agreed then and there to stay quiet about the discovery and not tell upper management. There would be no mention of the Easter Egg whatsoever and no acknowledgement of its existence from anyone at Tandy.

The CoCo 3 was saved; it would live to fight another day.

Siegel felt that Microware's action jeopardized the project and put Tandy in a precarious position. Microsoft was already being openly challenged by having its BASIC ROM patched by the programmers at Microware. Having the Easter Egg embedded in the ROM was the equivalent of Microware sticking the proverbial middle finger in the air, with Tandy providing the supporting hand and arm.

Tandy had reason for concern since it had far more to lose than Microware. The former's critical relationship with Microsoft, who up to that point had licensed its MS-DOS operating system for use on Tandy's PC compatibles and related machines, could have been in real jeopardy if this secret got out.

Siegel resolved he was not going to let that happen. In a scene reminiscent of a mob movie, he called Kaplan and made it clear that Tandy expected the people at Microware to never talk about the Easter Egg. According to Siegel, the expectation was kept in place for years. No one at Tandy or Microware talked about or acknowledged the Easter Egg during the remainder of the CoCo 3's shelf life.

Unfortunately for those involved, there was no way to stop people from learning about the secret keystroke, nor stopping pranksters in Radio Shacks from using the combination to put CoCo 3s used for in-store demonstrations into the "three amigos" mode. For a time, this kept many flustered Radio Shack managers busy trying to figure out just how this kept happening to their machines.

In time, the doors on the code of silence would bust open. Some 25 years after the fact, former president of Microware Systems Corporation Ken Kaplan would admit to personally coming up with the idea of the Easter Egg. It was, as he put it, a way of giving the brilliant engineers who worked so hard on the project a little credit. Mark Hawkins would echo Kaplan's sentiments on the Easter Egg and go on to further state that by creating the picture Microware actually fulfilled Tandy's original directive to fill the extra ROM area with random junk. By encoding a picture into the ROM, that is exactly what they did.

Without a doubt, the picture of the Three Mugateers was the most elaborate and ingenious hidden gem found on an 8-bit computer. Its inclusion added to the lore and mystery of the Color Computer 3 and remains a beloved quirk to this day as well as an easy way to perform a cold restart of the machine itself. Another little known fact was that the picture was a composite of three different photos captured with a DS-69B digitizer (Figures 8.2 to 8.4). Even more curious is the fact that Mark Hawkins and Todd Earles were actually wearing the exact same jacket in the photo, and the Microware logo was added in later with photo editing software on the Color Computer.

Siegel would later recall that competing interests and resources threatened the very existence of the CoCo. The machine that was used and loved by many actually remained in constant peril at Tandy, much like the sword that hung by a hair over the head of Damocles as he sat on the throne of Dionysius II of Syracuse. For

Figure 8.2

Original scan of Mark Hawkins that was used in the infamous Easter Egg. (Courtesy of Allen C. Huffman, who found these images on old Microware 5.25" floppy disks.)

Figure 8.3

Original scan of Tim Harris that was used in the infamous Easter Egg. (Courtesy of Allen C. Huffman, who found these images on old Microware 5.25" floppy disks.)

a short period in the summer of 1986, the hair that held the CoCo 3's own sword came dangerously close to breaking. Had it done so, the Color Computer 2 would have been the last in the series.

Yet, despite its precarious position, the Color Computer line managed to survive as it always had. And it would continue to defy the odds and survive for the years still to come.

With the CoCo 3's debut in Radio Shack stores in time for the 1986 holiday season, interest began to turn to software. Although the CoCo 3 could run most of the software available for its predecessor, the CoCo 2, customers were asking for applications that took advantage of the enhanced graphics and memory of the new machine.

Figure 8.4

Original scan of Todd Earles that was used in the infamous Easter Egg. (Courtesy of Allen C. Huffman, who found these images on old Microware 5.25″ floppy disks.)

As 1987 came, a number of new software titles for the machine began to emerge. On the productivity side, there was Tandy's own *DeskMate 3*, which featured a word processor, spreadsheet, database, and paint application. Although not file-compatible with *DeskMate* for Tandy's IBM-compatible offerings, the software was still viewed as a powerful productivity suite, and took advantage of the CoCo 3's new graphics and memory capabilities. It would be one of the first applications offered by Tandy that ran under OS-9 Level Two and carried on the legacy of *DeskMate* for the Color Computer, which ran under OS-9 Level One.

The fact that *DeskMate* ran under OS-9 at all was a battle in and of itself, according to Mark Siegel. An in-house software team at Tandy Electronics was tasked with writing *DeskMate 3* but felt restricted by OS-9's graphics APIs. Siegel insisted that it be done under OS-9 and adhere to specific graphic calls provided by the operating system. In the end, the OS-9 approach won out.

Another productivity program available for 1987 was *Phantom Graph*. Developed for Tandy by third-party contract developer Greg Zumwalt, *Phantom Graph* allowed for the creation of charts and graphs from imported data sets from popular spreadsheet programs like *Dynacalc*. On the entertainment side, *Zone Runner* (another title from Zumwalt), a space trading and combat simulation, and Epyx's *Rogue*, an official version of the popular procedurally generated dungeon crawling game that could take advantage of the 512K RAM expansion, were introduced as disk-based games for the CoCo 3 and also ran under OS-9 Level Two.

Within CoCo community circles, there was a huge amount of excitement for Tandy's newest Color Computer. After years of waiting, the boys at Fort Worth had finally created a machine that could compete more favorably with the Apples, Ataris, and Commodores of the day as well as get a head start on the coming software that would take advantage of its expanded features. CoCo fans had something genuinely exciting to put under their Christmas trees (or equivalents) in 1986.

The new year, 1987, started big for the CoCo 3. Unshackled from the 256 × 192 4-color graphical limitation of its predecessor, the CoCo 3's 320 × 192 resolution with 16 simultaneous onscreen colors opened the door to the possibility of a whole new range of software ports. Tandy's Mark Siegel set about to do just that, contacting companies like Sierra On-Line, Lucasfilm Games, Activision, Epyx, and other top software publishers at the time in an effort to convince them to bring their best titles over to the CoCo 3.

It was not a hard sell, as Siegel recalls. "The buying power of over 6,000 Radio Shack company and dealer stores was immensely compelling." Added to that, each store was expected to carry at least five copies of a single software title, giving the software companies an immediate and guaranteed order of approximately 30,000 copies, which were big numbers back then.

When approached about porting software to the CoCo 3, software houses would not even blink, Siegel recalled. "Their first question was: 'How many copies do you want?' Their second question was: 'When can you start selling it?'" As a result of Siegel's advocacy, ROM paks taking advantage of the new machine soon began to appear in Radio Shack stores everywhere, including famous cross-platform titles like Activision's puzzle game *Shanghai*, Sierra's action shooter *Silpheed*, and Epyx's action platformer *Rad Warrior*.

Disk-based game software running under OS-9 Level Two also began to arrive on the platform, spurring sales of Tandy's new FD-502 disk drive system, a $299 item. Top titles like subLOGIC's sophisticated *Flight Simulator II* were sold in Radio Shack stores, as was Broderbund's educational *Where in the World Is Carmen Sandiego?*, Activision's educational *Laser Surgeon: The Microscopic Mission*, Epyx's action-packed *Sub Battle Simulator*, and two famous action titles from Lucasfilm Games in partnership with Epyx: *Koronis Rift* and *Rescue on Fractalus!*

Yet perhaps the software products that arguably most legitimized the CoCo 3 as a competitive gaming platform for its day were the ports of Sierra On-Line's two most popular graphic adventures in 1988: *King's Quest III: To Heir Is Human* and *Leisure Suit Larry in the Land of the Lounge Lizards* (Figures 8.5 to 8.8). These games were exceptions in that they required the CoCo 3 to be equipped with 512K of RAM,* thereby limiting the target audience to those with better equipped systems. Nevertheless, the graphical quality, colors, animations, and

Figure 8.5

The title screen from *King's Quest III: To Heir Is Human.*

---

* The only other 8-bit computer version of these games was for the Apple II, which only required 128K, but did not have the same multitasking capabilities.

Figure 8.6

A scene from *King's Quest III: To Heir Is Human.*

Figure 8.7

The intro screen from *Leisure Suit Larry in the Land of the Lounge Lizards.*

Figure 8.8

A sample scene from *Leisure Suit Larry in the Land of the Lounge Lizards.*

scope of these games were worth it. An even more impressive feat was that the games themselves ran under OS-9 Level Two, though tricks were employed to gain access to the needed RAM in somewhat unconventional ways. Still, it was possible to play one of the games in a window, while switching to another window and running other programs simultaneously.

On the software front, things were looking rosy indeed for the Color Computer 3.

Tandy's push for OS-9 Level Two as the basis for applications and games was an intentional strategy aimed at making the CoCo 3 a more serious platform. Yet the lack of a standardized graphical user interface was a glaring omission in an otherwise growing ecosystem of software. Multi-Vue was created to change that.

Originally given to Microware as a specification from Tandy in 1986, Multi-Vue was envisioned as a true point-and-click graphical user interface like that found on Apple's Macintosh computer. With this specification in hand, Microware began the monumental effort to code up the graphical "shell," as well as the underlying system support software that would make up the Multi-Vue platform. In 1987, Multi-Vue began appearing at Radio Shack stores bundled with several sample applications, including a clock and a calendar.

Another glaring omission on the CoCo 3 was a joystick/mouse interface capable of addressing the 640 × 192 graphical screen that could now be attained. With joystick and mouse values only going from 0 to 63 in both the X and Y directions, fine mouse cursor movements were impossible to attain, even with the new two-button white Deluxe Color Mouse that Tandy had released.

The Hi-Resolution Joystick Interface was designed to address the problem. Housed in a small rectangular box, the interface plugged into the CoCo 3's joystick and cassette ports; a mouse or joystick would then plug into it (Figure 8.9). Use of this interface increased the effective resolution of the joystick ports 10-fold, allowing finer grained mouse-driven cursor control.

In an interesting strategic move, Tandy passed this interface design through CoCo programmer Steve Bjork to refine and finish. As Siegel noted ironically, it was easier for a third-party designer to sell something to Tandy than to have engineers from within Tandy sell the design to their buyer. Having Bjork finish the product and pitch it to Tandy was a roundabout way of getting the product on Radio Shack shelves. Tandy flew Bjork to Fort Worth, and then paid him royalties. In two weeks, there were two prototypes. One went to the factory for pricing analysis, and the other went to Microware for OS-9 support.

Although Tandy's software offerings for the new Color Computer 3 were certainly nothing to sneeze at, notable titles also came from outside of Tandy Towers' purview. And it was not surprising that, given the CoCo 3's advanced graphics, the most eagerly anticipated software offerings would focus on the new computer's capabilities in this area.

The June 1987 issue of *THE RAINBOW* magazine featured a tantalizing debut ad for an exciting new graphics drawing program called *Color Max 3* (Figure 8.10). The program built upon the palette-styled tool menu of drawing programs that were available at the time for the Macintosh, featuring a similar look and feel to that of Colorware's earlier *CoCo Max II* drawing application for the CoCo 2. Unlike that product, however, *Color Max 3* took full advantage of the enhanced resolution and color space of the CoCo 3.

Created by Erik Gavriluk and Greg Miller, and sold by Computize, *Color Max 3* was the first CoCo 3-specific graphics drawing program, and appropriately took the community by storm. The July 1987 full-page ad featured gorgeous screenshots of computerized art and a rainbow spilling out of a CM-8 monitor

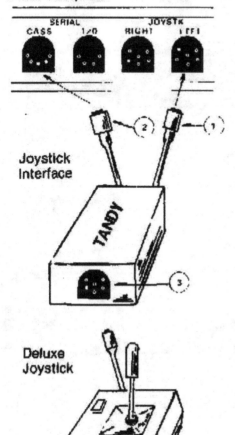

Color Computer

SERIAL   JOYSTK
CASS   I/O   RIGHT   LEFT

**Joystick
Interface**

TANDY

**Deluxe
Joystick**

INSTALLATION

Before connecting the Interface, be sure that power to the Color Computer and peripherals is turned off.
1. Connect the 6pm DIN connector 1 of the Interface to one of the 2 JOYSTICK lacks of the Color Computer.
2. Connect the 5pin DIN connector 2 of the Interface to the CASS jack of the Color Computer.
3. Connect the 8-pin DIN connector of the Deluxe Joystick to the jack 3 on the Interface.
OPERATION

Your joystick or color mouse will operate just as if were connected directly to the Color Computer. Note, however, that the
resolution on the display will be 10 times higher than normal.
To begin using the Color Computer with the Interface, load the appropriate program and RUN it.

Figure 8.9

A page from the Tandy Hi-Resolution Joystick Interface Owner's Manual showing
how to hook everything up.

Figure 8.10

The advertisement introducing *Color Max 3* in the June 1987 issue of *THE RAINBOW*.

and onto the keyboard of the CoCo 3 below, while a pot of gold coins sat regally on the screen. The disk-based software required a 128K CoCo 3, a mouse or joystick, and, not surprisingly, Tandy's Hi-Resolution Joystick Interface.

Gavriluk recalls the start of *Color Max 3*'s development: "Tandy announced the CoCo 3 on July 30th, 1986, but Greg and I didn't receive advance machines.

We got ours at Radio Shack sometime in October, the first day they arrived in the stores. That night, Greg had the GIME chip doing perverse things and I was busy seeing how fast I could make zoomed 2X ellipses fill in with different colors and patterns.... On April 10th, 1987, we debuted *Color Max 3* at the Chicago RAINBOWfest. *Color Max 3* was a thorough, complete, and surprisingly bug-free product. It beat *CoCo Max III* to market by six months."[*]

Not only did it make CoCo 3 owners drool, it was also seen as a shot across Colorware's bow. Colorware was a steadfast advertiser in *THE RAINBOW* and multipage ads for *CoCo Max II* graced issue after issue of the magazine. Now with *Color Max 3*'s announcement, *CoCo Max II* was upstaged. Colorware's John Monin, who continued to run advertisements for *CoCo Max II* in *THE RAINBOW* on a monthly basis, did not take the threat sitting down, however. Instead, he recruited a prolific young Canadian programmer named Dave Stampe.

Stampe started out as a self-taught programmer who disassembled and studied the *CoCo Max II* product that Monin sold through Colorware. Inspired by graphical algorithms and techniques, Stampe began to create his own small graphics programs, working on enhanced input resolution and drawing effects, even going as far as sending his work to Colorware. Monin saw the work and was interested in having Stampe apply his talents to creating a new graphics program for the CoCo 3. Stampe agreed and the collaboration began.

The young programmer took a novel approach to developing software by writing his own assembler and disassembler tools. He also had a heavily modified disk operating system, which amounted to a true homebrew development platform. With his own tools in hand, he began to craft what would become *CoCo Max III*, all the while working at arm's length with Colorware. Suggestions went back and forth, but Stampe mostly worked in isolation. At times he would ask for and receive special equipment from Colorware, like an ink jet printer that he needed to write a driver for. Other than receiving financial support in the form of royalties and some upfront stipends, there was no direct intervention or micromanagement from Colorware.

The *CoCo Max III* product was an eight-month stint, as Stampe recalls. "It was a starter project, and my first major programming effort." When it was complete, Colorware announced it to the world in a full page ad in the September 1987 issue of *THE RAINBOW* (Figure 8.11). Priced at $79.95, *CoCo Max III* offered a number of impressive drawing features that ran on the stock 128K CoCo 3.

Coincidentally, that same September 1987 issue of *THE RAINBOW* featured a third graphics package for the CoCo 3 known as *The Rat*, from the Canadian company Diecom Products, right on the inside front cover. It seemed the war of the graphics programs was being waged in earnest.

CoCo Max III used an interesting form of hardware-based copy protection. The Tandy Hi-Resolution Joystick Interface, which increased the effective horizontal and vertical resolution of a standard joystick or mouse, was modified by Colorware (a design that Stampe also contributed) so as to not

---

[*] Nickolas Marentes, *CoCoNUTS: Interviews with People Who Helped Shape the CoCo* (1998–1999).

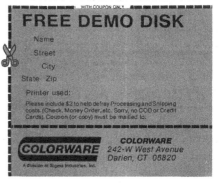
Figure 8.11

The advertisement introducing *CoCo Max III* in the September 1987 issue of *THE RAINBOW*.

require the second connector to be plugged into the cassette interface port. Instead, the modified adapter sported a single connector to the joystick port, and the software was set to detect this specific modification. This meant that copies of *CoCo Max III* were useless without the hacked adapter. *Color Max 3*'s dependency on Tandy's unmodified Hi-Res adapter made it easier to copy and thus more susceptible to piracy.

The competition came to a head in November 1987 with Colorware putting out a full page head-to-head comparison between its *CoCo Max III* and Computize's *Color Max 3* in that month's issue of *THE RAINBOW*. It worked. Throughout the remainder of that year and into 1988, Colorware's monthly advertising blitz totally overshadowed *Color Max 3*. Even an updated release of *Color Max 3*, known as *Color Max Deluxe*, could not stop the momentum of Monin's graphical gem.

According to its coauthor Greg Miller, *Color Max 3* "was overshadowed primarily because of advertising. Erik and I worked with a couple of great guys who ran a company called Computize. They made it clear to us from the beginning that they didn't want to play 'advertising war,' and alluded more than once to the notion that Lonnie Falk [the publisher of *THE RAINBOW*] was notorious for heavily promoting products of his friends, while offering substantial advertising discounts. I can't say if any of that is true, but Bruce and Kenny [the Computize folks] felt that it would be a losing battle."[*]

Colorware did not stop with *CoCo Max III*. Monin continued to dominate the graphics market niche on the CoCo 3, commissioning Stampe to write yet another software hit, an impressive WYSIWYG (What You See Is What You Get) desktop publishing application called *Max-10*, which debuted in June 1988. Like *CoCo Max III*, *Max-10* employed hardware-based copy protection but with a device called a "clicker," which plugged into the cassette port of the CoCo 3. Without it, the software would not fully load and run.

Up to that point, word processing on the CoCo consisted of editing on a text screen and printing straight to a printer. The concept of a word processor with true WYSIWYG was still considered unattainable by most Color Computer developers. Monin challenged Stampe to prove the naysayers wrong. After all, this was the guy who did *CoCo Max III*.

Stampe worked for a year developing *Max-10*. "It was a much bigger effort [than *CoCo Max III*]," he recalled years later. To get the feel of how the program should operate, John Monin sent Stampe a Macintosh loaded with software. Stampe used the operation of the *MacWrite* word processor as a design template for *Max-10*. The fonts used by the program were partially created and edited from public domain fonts. "Dot matrix printers made it a challenge to properly print fonts," Stampe remembers. "I wrote a BASIC program to help with font creation."

Colorware promoted the *CoCo Max III* and *Max-10* combination in its advertising in *THE RAINBOW*, and the products continued to sell well into the early 1990s. Stampe points out proudly that the money that he made from *CoCo Max III* and *Max-10* was used to pay his way through university. Unfortunately, for everyone ranging from those who made a living from the CoCo platform to its casual fans, the good times would soon be heading into their twilight years.

---

[*] Marentes, *CoCoNUTS*.

9

Made in the USA

Mailed out around Labor Day of each year to customers, Radio Shack's annual catalog featured both familiar items as well as new products that would appear the following year. In the fall of 1988, CoCo 3 owners who received their 1989 Radio Shack Catalog were treated with a slew of major new cartridge-based games that were coming out, including *Super Pitfall*, *GFL Championship Football II*, *Castle of Tharoggad*, *Silpheed*, *Tetris*, and *Soko-Ban*. New disk-based software offerings were light in comparison, leading many to speculate that Tandy was transitioning the Color Computer 3 away from a home computer and more toward a game machine. There was one new disk-based game title, however, on page 166 of the same catalog called *The Last Ninja*.

The description was tantalizing and gave a clue to the technical breadth and programming challenges that the game would cover: "Scale forbidding mountains, cross lakes and rivers. Meet an array of formidable foes, human and otherwise. Master new and dangerous tasks—like Ninja magic. Because...only you—the last Ninja—can avenge the brotherhood and recapture the secrets of the Ninja."

The game, which was already available for and something of a legend on the Commodore 64, was an immersive, third-person, isometric perspective action adventure that followed a black-clad ninja through detailed outdoor scenery. Tandy and Activision agreed to terms that would allow the game to be ported to the Color Computer 3 under OS-9 Level Two. All that was needed was an engineer to do the work. Initially, Tandy approached longtime programmer, Dale Lear, who turned down the work. A second CoCo developer, Rick Adams, was contacted and asked to work on the game. Adams already had an impressive body of CoCo work, including the popular *Temple of ROM* and *Shanghai* game cartridges. He expressed interest in porting *The Last Ninja* to the CoCo 3, buoyed by a tidy $2,000 signing bonus.

The mandate was clear: port the game to the CoCo 3. If the deadline could not be met, Adams was to give enough lead time to salvage the project. "I began to work on it," Adams remembers. "It was a sophisticated game. There was a crazy deadline ... four, maybe six months." As Adams recalled years later, the warning signs were there early on: no clear specifications, nebulous ideas about gameplay, and a bundle of raw graphic data but no information on what levels would be required.

After months of working into the early morning hours, Adams began to come to the realization that he may have bitten off more than he could chew. Aware of his obligations, he notified Activision that the game would not be completed in time. Still keenly interested in making the game work, both Tandy and Activision decided to hand off the project to game programmer Steve Bjork. Adams would keep the $2,000 signing bonus that he received but would not get any royalties from the game when it sold.

Excited at the opportunity to turn the fortunes of the game around, Bjork met with Adams at Activision's headquarters for the hand-off of the source code. As both programmers looked over the work at that point, Bjork indicated areas of the code where he felt he could make improvements and optimizations. But in the end, not even one of the Color Computer's top programmers could complete the game on time.

Adams now believes that the project cratered for several reasons, including technical challenges, a lack of specifics, and the sheer size and scope of the project against the time constraints given. "The project was just too big for one person to do it alone with the aggressive schedule required." As an example of the wild expectations that were placed on the game, he recounts: "There was one level that involved hopping over stones in a river, but what nobody bothered to tell Steve or I was that a freaking *dragon* came out of the river halfway over to do battle with you. Steve was kind of disturbed that nobody explained this little detail to him. I didn't get far enough to figure this out, but I'd have to say I agree with him."

Looking back on the experience years later, Adams was both reflective and diplomatic: "It was an unhappy time for all concerned." Years would pass after the game was canceled when, in a twist of fate, Bjork would bring a copy of the unfinished game to the 2007 "Last" Chicago CoCoFEST and demo it to the audience there. Even in its incomplete state, it made for an impressive show.

Nothing was viewed more sacrosanct at Radio Shack than accuracy in the content of their annual catalog. If the catalog advertised it, customers read it and expected the product to be available. As is the case with any developing product, software offerings are planned and scoped, then take some time to write, test, and package. It was not uncommon, in certain situations, to "preannounce" the availability of an upcoming product.

Tandy's Mark Siegel remembers taking a bit of heat for the product's cancellation. "I had to explain that the product wasn't going to make it to [then Radio Shack President] Bernie Appel since it was in the catalog."

Appel's response was simple and direct: "Don't let it happen again."

Luckily, other areas within the CoCo 3 were not nearly as negative. For instance, no sooner had OS-9 Level Two been released in early 1987, than an intrepid and skillful OS-9 hacker named Kevin Darling began to disassemble and comment on various components of the operating system. The brute-force exercise, often relegated to the hard-core, bulldog geek types, gave Darling a deep understanding of the internals of OS-9, and the effort led Darling to write his popular book, *Inside OS-9 Level II*, in 1987, which explained the inner-workings of OS-9 on the Color Computer 3.

Yet his tinkering did not stop there. Darling began to radically rewrite certain parts of the operating system, especially the components that dealt with graphics and the windowing system. For several years, CoCo 3 users who were also strong OS-9 aficionados longed to see an upgrade to the work that Microware and Tandy did, both in the OS-9 operating system itself and with Tandy's flagship GUI for the CoCo 3, Multi-Vue. Although OS-9 Level Two had windows, and Multi-Vue exploited them, they were still bound by certain limitations: windows, once defined with a fixed width and height, could not be resized or moved from their position. CoCo 3 and OS-9 power users really wanted truly resizable, movable windows.

What many did not know at the time was that Darling and a handful of other programmers were working on enhancements to do just that, along with improvements to other aspects of the OS-9 operating system (Figure 9.1). Not only that,

Figure 9.1

Kevin Darling, Kent Meyers, Dale Puckett, and others gather around a CoCo 3 in a hotel room during the 1988 Chicago RAINBOWfest. (Courtesy of Ron Lammardo.)

but Darling's secret group was in touch with people at Tandy and Microware to actually bring the work together into an upgrade that could be offered via Tandy's Express Order Software mail-order service. Contracts were signed and lips were mostly sealed as the work continued.

Several parts of the operating system came under the team's microscope for improvement. Darling and others combed the kernel source for optimizations and enhancements. Other prime targets were OS-9 modules responsible for performing primitive graphic drawing operations, which yielded the most visible improvements. Graphic windows finally became resizable and movable.

Utilities were also improved. Team member Ron Lammardo took the venerable OS-9 shell under his wing, expanding its functionality to include basic scripting commands and variables as well as additional built-in commands. Other OS-9 notables such as Chris Burke, Bill Dickhaus, Mark Griffith, and Bruce Isted made various contributions to everything from the pipe file manager to new system modules to managing large blocks of memory via system calls.

Even Multi-Vue was placed under scrutiny for improvement. Fellow OS-9 programmer Kent Meyers took on the task of disassembling the package's various code modules. Even more impressive was that Multi-Vue was mostly written in the C programming language, and the resulting binary files were larger and more verbose, making the disassembly and commenting process more challenging. Meyers's work brought significant speedups to the graphical user interface.

Despite all of the improvements to the OS-9 Level Two package, in the end, Tandy decided not to market the upgrade. As a result, work on the project was suspended. Yet, instead of totally abandoning the work, Darling decided to release small pieces and parts, as exemplified in the 1989 GrfDrv patch, which he provided to the CoCo community as a Christmas holiday present that year, disseminating the patch over the Delphi and CompuServe online networks. This enhancement alone increased graphics performance on the CoCo 3 under OS-9 Level Two considerably. Other parts of the upgrade would eventually leak out, but the entire package remained under wraps for years. It would be a crusade of sorts, led by of all people, an Anglican monk, to bring the mysterious upgrade out of the dark to fully see the light of day.

Even though Tandy was not exactly leading the way for new CoCo 3 initiatives, the platform was not necessarily being ignored by them either. After years of CoCo 2s, and, later, CoCo 3s being made in Tandy's factory in South Korea, the company began looking at bringing the manufacturing of its latest Color Computer back to U.S. shores. The CoCo 3 was specifically mentioned in a November 22, 1987, article in *The Dallas Morning News* titled "U.S. Companies Are Coming Home."

"Beginning next year, Tandy will manufacture its Color Computer 3 at two plants in Ft. Worth after four years of production in Seoul. 'All of the production we moved in the past moved from the United States to the Orient,' said Tandy Chairman John Roach. 'This is the first time it's been from the Orient to the United States.'"

The article continued: "'When a new computer plant is finished in July, 1988, Tandy's employment in Fort Worth may be up by as many as 500 workers. The ripples could mean jobs elsewhere, too. Circuit boards, molded-plastic cabinets and components, obtained in the Orient for the Color Computer 3 made in Seoul, will next year come from Tandy facilities and suppliers in the U.S.,' Roach said."

The CoCo 3 began production in the fall of 1988 at Tandy's Fort Worth factory. CoCo 3s were now "Made In America" again. And none too soon. That summer, Tandy's factory in South Korea was the subject of a labor dispute between male and female workers. According to news reports, the males, frustrated at their female counterparts' protests, unceremoniously strung the ladies upside down by their feet, then proceeded to beat their bare feet with canes. It was an embarrassing moment for Tandy, which worked quickly to address the issue, but forced production of some other products to move to Taiwan.

At the new factory, CoCo 3 motherboards would be stuffed with parts and go through a wave soldering process to "weld" the chips into place. Although these units were functionally the same as their South Korean counterparts, there were a few obvious telltale signs of their origin. American CoCo 3s had badges with a slightly larger TANDY text on the top, as well as half spherical, light gray rubber feet that were glued to the bottom of the case. In contrast, the CoCo 3s made in South Korea several years earlier used black, flat hard plastic feet that were pinned through a hole in the bottom of the case.

The changes were just a part of several that were done to reduce manufacturing costs, something that Tandy was obsessive about. And they certainly had an ally in Kenji Nishakawa, Tandy's cost-hawk who ran the CoCo 3 manufacturing facility (as well as the older CoCo 2 factory) in what was an old Piggly Wiggly supermarket on Blue Mound Road.

A Japanese native, Nishakawa was legendary for his ability to put the squeeze on costs, finding areas to save both money and improve the manufacturing process. His use of lower-waged Thai, Cambodian, Laotian, and Vietnamese female workers on the factory assembly line helped to keep costs down, as did his odd choice of factory equipment. When it came time to clean the CoCo 3 motherboards that were coming down the assembly line, Nishakawa chose to bypass industrial cleaning systems, instead opting for cheaper consumer dishwashing machines to run the boards through.

When a discoloration on the connectors on the back of the CoCo 3 came to light due to the soldering process, Kenji found an interesting solution to bring the connectors back to their pristine black color. He touched the side of his nose, and then rubbed his finger on the distressed connector to find that the original black color returned. "Aah, nose oil!" he exclaimed. Although linseed oil was actually applied from that point forward, the adage stuck. From then on, solutions to problems within Tandy were referred to jokingly as "nose oil."

However, Nishakawa's aggressive cost-saving measures did go too far one time. In an effort to save some pennies, a part substitution was made that caused some of the CoCo 3 video colors to wash out. As would be expected, Tandy's quality control quickly caught the problem. Nishakawa made the switch back to the old part, losing his cost savings but in turn saving the CoCo 3s "colorful" reputation.

Nishakawa was not the only person pushing the envelope for the CoCo 3, however. Even as Tandy's focus for the CoCo 3 appeared to change from a serious disk-based system to an entertainment console, the CoCo community continued to expand on what the tiny computer could do. Nothing was more emblematic of that progress than the addition of inexpensive hard drives to the CoCo 3.

Compared to floppy systems sold by Tandy and other vendors, hard drives offered a wealth of storage space, as well as much higher data loading and saving speeds. Prices tended to keep these elite performers out of the reach of most CoCo owners, but an intrepid idea from a Motorola employee in 1988 led to a hard-drive revolution in the Color Computer community. *THE RAINBOW* magazine declared 1988 the "Year of the Hard Disk," in large part thanks to the Burke & Burke Hard Drive System.

The system's creator and namesake was Chris Burke, a Motorola employee and a CoCo user since the days of the very first Color Computer (Figure 9.2). Burke's involvement in the CoCo started in 1980 while he worked as an intern at Pellerin-Milnor Corporation in Kenner, Louisiana. The company's small electrical engineering department had taken on the young engineering student from the University of Wisconsin at Madison, giving him the chance to work with interesting hardware like pressure transducers and sonars. Tasked with acquiring a computer to perform the control of a prototype system, Burke's boss, Dan Garcia, went to a local Radio Shack and bought a 4K Color Computer to perform control operations. In no time, Burke was hooked.

After graduating from college with a degree in electrical engineering, Burke took a job with Motorola. By this time, he had expanded his CoCo, subscribed to *THE RAINBOW* magazine, started running OS-9, and began writing software for himself. He also began to look for a hard drive solution but was quickly discouraged with what he found.

Figure 9.2

Chris Burke working on his CoCo 3, circa 1986. (Courtesy of Chris Burke.)

As Burke searched, he realized that the existing offerings were expensive and mostly piecemeal solutions. Tandy's own hard drive controller for the Color Computer was $129, and that did not even include the hard drive or the power supply. It also interfaced with older technology even for the time: SASI hard drives.

Far from being dissuaded, the Motorola engineer began to examine the possibility of interfacing hard drives to the Color Computer less expensively. When *THE RAINBOW* magazine's regular monthly contributor Marty Goodman opined that someone should look into then-popular MFM and RLL drives of the time, mostly found on PCs, Burke took notice. Examining the Western Digital interface board, he began to design an adapter that would physically mount the PC-based ISA bus card to the 40-pin edge connector of the Color Computer, dubbed the "CoCo XT II" (Figure 9.3). Burke even found room to place a real-time clock on the board.

As the design went through various phases of testing and vetting, Burke envisioned selling turnkey systems: the adapter, hard drives, and case as a single system. Both he and his wife Trisha began the work of soldering, assembling, and testing boards right in their apartment. And thus, Burke & Burke was born.

Shortly after placing an advertisement in *THE RAINBOW* magazine for the Burke & Burke Hard Drive System, orders began trickling in. CoCo owners saw the relatively cheap and accessible PC-clone hard drives as perfect add-ons to their systems. Burke also sold Hyper I/O, a hard drive accessible version of the CoCo's Disk BASIC, as well as drivers for OS-9 users. Now, every CoCo owner could enjoy the full speed and storage that was up to that point much more expensive to obtain.

Before long, Chris Burke's new product started to catch fire. Throngs of folks mobbed his booth at RAINBOWfests, clamoring for the now-famous hard drive system. It did not hurt matters that Burke had a little help. His wife, Trisha, a rather attractive United Airlines flight attendant, manned the booth right alongside her husband and knew enough about the products to answer most questions from the geeky, male-dominated attendees. Often, they would approach the booth timidly but eagerly with their questions, and end up leaving with hard drive systems under their arms. Her presence at the booth unquestionably added to Burke & Burke's success.

In addition to the Burke & Burke Hard Drive System, other offerings began to appear. Joe Scinta of Ken-Ton Electronics out of Tonawanda, New York, designed and began marketing an impressive hard drive controller that interfaced with ever-popular SCSI hard drives, which were known for their fast access times, durability, and significant capacities. Scinta paired his hard drive system offering with Roger Krupski's RGB-DOS, an enhancement to Disk BASIC, which allowed very effective and compatible hard drive access to existing CoCo software. Owl-Ware also started offering a SCSI-based hard drive package and BASIC/OS-9 software support as well.

Although the hard drive system was Burke & Burke's mainstay, Chris Burke had other hardware and software offerings, mostly aimed at the serious

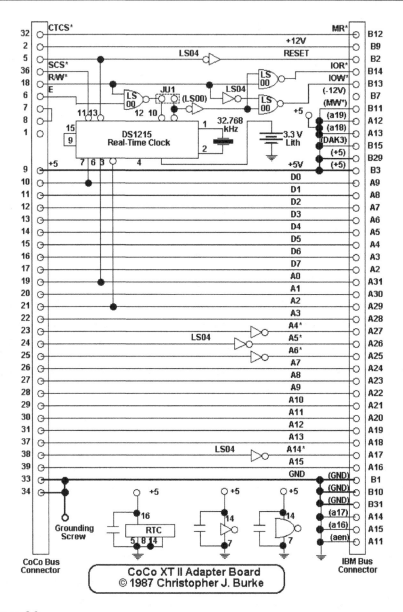

Figure 9.3

The Burke & Burke CoCo XT II schematic. (Courtesy of Chris Burke.)

CoCo users who used OS-9: a voice synthesizer, a file system utilities package, and a conversion of Tandy's *Cyrus* chess program pak to run under OS-9.

As the business grew, so did Chris Burke's stature in the community. In 1991, he was invited to be the keynote speaker at the Chicago RAINBOWfest, an honor reserved for very few notables within the Color Computer community.

Even with the success of Burke & Burke's hard drive system, Chris Burke had one more idea: replacing the venerable but aging 6809 microprocessor in the CoCo 3 with a more powerful Motorola chip: the 68306. Dubbed "The Rocket," the product was considered a major upgrade that would propel CoCo owners to a new level of usability and possibilities. The Rocket would run Microware's OS-9/68K operating system, a natural successor to the OS-9 Level Two that Tandy had commissioned. Despite having built a prototype as a working proof-of-concept, in the end, the product failed to launch, as Burke could not get a sufficient number of units presold to overcome Microware's licensing costs.

Despite the nonlaunch of The Rocket, Burke & Burke ended on a good note, closing its doors in 1992, knowing that it served the CoCo community well and was compensated sufficiently. As Chris Burke recalled years later, "I never got rich off of Burke & Burke. It was never even a full-time job—I was at Motorola in those years. The money we earned at Burke & Burke went back into the business, and into savings. Eventually, it became the American Dream: a down payment on a house."

As the CoCo 3 entered its third year, there was no shortage of eager fans who continued to find ways to push the machine to new and almost unimaginable limits. Evidence of such work splattered the pages of *THE RAINBOW*'s advertisements. One example of this was a novel sound-recording application from a company called GIMMESoft. Its 1988 offering, *Maxsound,* brought primitive but effective audio digitizing to the CoCo 3 through the joystick port's digital-to-analog converter.

Another impressive program on the sound and music front was OS-9-based *UltiMusE,* from Mike Knudsen. It was developed for the CoCo 1 and 2, but performed even better on the CoCo 3 thanks to its additional 64K of RAM being available for the program, graphics images, and I/O buffers. The user got access to the memory left over for a score file that maxed out at an impressive 1500 notes and other score objects. Released successfully as shareware to gauge the CoCo community's interest in musical usage of the new MIDI standard, donations and moral support poured in, and some incredible musical arrangements were crammed into the 1500 notes.

The mouse and 256×192 monochrome graphics screen were used only for editing notes on *UltiMusE*'s Score Screen. All the menus were 32×16 character hardware text-screen listings to remind the user what keys to hit for the commands. Even setup and layout was strictly textual. The last and most powerful version of *UltiMusE, UltiMusE III,* was commercially released in 1988 with a $54.95 price tag (Figure 9.4). It featured full CoCo 3 support with much better 640×192 graphics, user-selectable colors, and a new graphical layout screen. It amounted to an impressive use of the OS-9 Level Two operating system.

"Soon after, I developed virtual memory for the score data, allowing up to 32,767 objects in up to 32 pages, though the 6809's address space was still 64K. This was totally transparent to the user, and since all pages resided in the 512K RAM, very fast," recalls Knudsen.

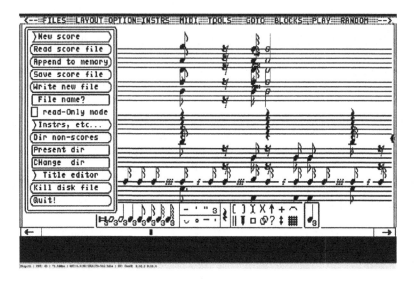

Figure 9.4

A screenshot of *UltiMusE III*, the phenomenal CoCo 3 music program. (Courtesy of Bill Pierce.)

Knudsen continued. "As more features were added, I used another 8K block to map in program subroutine modules from separate disk files, thus extending virtual memory to the program, as well as data. The most recent version comprises *UltiMusE III*, Fran, and nine 8K modules. Today, these modules would be called DLLs, but OS-9 supported these back when Bill Gates was still debugging QBasic."

Then there were those who seemed able to push the CoCo 3 to its absolute limits. An example of such extreme programming came in 1989, when a young Canadian CoCo 3 owner named John Kowalski released a program on his BBS that took the CoCo world by storm. The deceptively simple name of DEMO1 was actually an impressive demonstration of both the range of colors that the CoCo 3 could produce on the screen at one time and the scrolling feats of the GIME chip.

Users who downloaded and ran the demo written by the "Sock Master" (Kowalski's *nom de guerre*), were treated to a smorgasbord of color and waving text with a scrolling banner across the bottom fifth of the screen. The program spread across the communication networks like wildfire and catapulted Kowalski into perpetual CoCo fame. Kowalski would later go on to create the equally impressive DEMO2 in 1992 and one year later BOINK (Figure 9.5), a rendition of the Commodore Amiga's famous bouncing ball demo, and among the best ever created for an 8-bit computer. It would prove a fitting tribute as the CoCo's mainstream commercial existence headed to its natural conclusion.

(a)

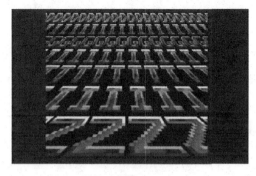

(b)

(c)

Figure 9.5

Screenshots of the impressive (a) DEMO1, (b) DEMO2, and (c) BOINK CoCo 3 demos. (Courtesy of John Kowalski.)

The Time Has Come

Like clockwork, Radio Shack's annual catalog appeared in September 1990 prominently showcasing two pages of CoCo 3 products and peripherals. The continued appearance of the Color Computer series in the general catalog's pages gave hope to CoCo owners that their favorite computer would still be sold in the upcoming holiday shopping season.

Just over a month later, however, on October 26, 1990, hopes were dashed when Tandy's Ed Juge announced through a press release that the Color Computer 3 would be dropped from its computer lineup. The news was devastating to fans of the platform but not totally unexpected, as rumors had been rampant for some time that such a move was in the works. It was no surprise that Tandy's focus would turn exclusively to its popular MS-DOS-based line of PC compatible computers, a class of systems that had pushed most other platforms to commercial irrelevance, if not total extinction. With the TRS-DOS-based TRS-80 Model 4D representing the final single-page appearance for that long-lived series in the 1990 Tandy Computer Catalog, the CoCo 3 would end its run as the last gasp for non-Microsoft-based desktop computers represented in its general catalog and still actively sold within Radio Shack stores (Figure 10.1).

Despite the lack of surprise, the news still came abruptly to Tandy's Mark Siegel and Color Computer buyer Barry Thompson, who got a call one day from

Figure 10.1

The 1991 Radio Shack Catalog showing the CoCo 3. It would be the last catalog to feature the computer. (Courtesy of www.RadioShackCatalogs.com.)

the Vice President of Computer Merchandising Graham Beachum. A Tandy newcomer from IBM, Beachum had already made waves by insisting that professional employees at the Tandy Towers wear red ties and remove all facial hair, an obvious homage to his former employer. Siegel and Thompson arrived at the office on cue and got the news delivered to them in a straightforward manner from the VP: the CoCo 3 was finished.

Siegel and Thompson were understandably disappointed, despite knowing that this day would eventually come. They attempted to appeal for clemency in a final, last ditch effort to prevent their nearly decade-long work from being shelved, but Beachum was unmoved.

What came next from the brash executive was perhaps the most insulting and degrading remark that the poor Color Computer ever received: "You can put perfume on a pig, but it's still a pig."

And just like that, the CoCo was unceremoniously dumped once and for all from Radio Shack's computer lineup.

Despite whispers of a prototype Color Computer 4 that supposedly existed in the confines of the Tandy Towers, Siegel squashed the rumor. "There never was a design for a CoCo 4. It never was a planned product. No work was put into one."

Although the trend appeared negative, online activity continued to spur interest in the Color Computer 3. Former Falsoft employee and OS-9 guru Greg Law recalls the intense discussions that occurred on the Delphi Information Service's CoCo and OS-9 forums at the time. An active participant himself, Law managed the forums under his Delphi handle "GREGL." New third-party software products were still being developed, people were still using their Color Computers, and OS-9 Level Two continued to give the CoCo 3 the lifeblood it needed to carry on.

Still, avid CoCo fans felt the window of opportunity closing on their beloved platform. As it now appeared, the much hoped for "CoCo 4" would not be coming from Radio Shack. Instead, a number of contenders would spring up to propel the Color Computer community into a new era of 16-bit goodness, riding on the back of Motorola's 68000 microprocessor, along with its crowned-king operating system, OS-9.

Paul K. Ward was no stranger to the CoCo scene. As a Color Computer user and author of the 1988 book *Mastering OS-9 on the Tandy Color Computer*, Ward had already gained credibility as an OS-9 user-expert. After bearing witness to what was fully expected to become the demise of the CoCo 3 at the hands of Tandy, he decided to act on bringing a worthy successor to the legion of Color Computer fans in the form of the MM/1, or MultiMedia One.

Designed by electrical engineer Kevin Pease, the MM/1 would be a new yet still somewhat familiar computing experience for Color Computer users. Based on the Philips/Signetics 68070, an instruction compatible version of the Motorola 68000, it would also contain as its flagship graphics processor the SCC66470, the same chip used in Philips's new CD-i multimedia player set top boxes. The MM/1 system would be stocked with 1MB of RAM, which could initially be expanded to 3MB, then 9MB via a backplane upgrade and second board.

Ward started Interactive Media Systems, or IMS, to bring his vision of a Color Computer successor to market. Considerable time and money were invested into marketing the MM/1: ads were placed in *THE RAINBOW*, brochures were printed, and a video was produced expounding the virtues and capabilities of the new machine. OS-9 programmer Kevin Darling was contracted to develop the windowing software based largely on Tandy's Multi-Vue graphical environment. Called K-Windows, the graphics package took advantage of the increased resolution of the graphics chip and provided partial and full-screen overlapping resizable windows, undoubtedly an inspiration from his work on the OS-9 Level Two Upgrade.

Unfortunately for IMS, storm clouds began to gather on the horizon. Sales went unfulfilled due to the lack of FCC approval, leading to a stall in much needed revenue for the small company. Demand for the product was clearly there, but confidence was beginning to wane due to the lack of availability. If something was not done quickly, a downward spiral of "no confidence/no sales" would spell doom. In an attempt to do an end-run around the foot dragging of the FCC, Ward began selling the MM/1 in kit form, which didn't require the coveted approval.

Meanwhile, another longtime Color Computer and OS-9 supporter, Frank Hogg, got into the action with not one but two potential CoCo successors: the Tomcat TC-9 and the TC-70. The TC-9 was Hogg's vision for a more compatible, truer successor to the CoCo 3. Based on the same 68B09E microprocessor as Tandy's Color Computer, the TC-9 system was designed by Bob Puppo, who also created the famous Puppo XT keyboard interface for the Color Computer.

The TC-9 motherboard accepted—and required—the same GIME chip that was found in the CoCo 3 (Figure 10.2), which meant either ordering the part from Tandy's National Parts ordering center or scavenging a GIME from an existing CoCo 3. Microware's CoCo 3 Super BASIC in ROM was also needed to bootstrap the computer to a functional state. A separate board contained serial, parallel, and joystick ports for communications and input/output (Figure 10.3).

The TC-70 was Hogg's answer to Interactive Media Systems' MM/1, sporting the same 68070 processor and graphics chip in a uniquely styled case. It also ran Kevin Darling's K-Windows software under the OS-9/68K v2.4 operating system.

The third and final contender was the 68000-based System IV from Delmar Company, out of Middleton, Delaware. Its owner, Ed Gresick, had been a longtime OS-9 user and Radio Shack franchise owner who was known for running his entire store's inventory and sales system on a CoCo 3 with OS-9 Level Two. A radical departure from the other offerings, the System IV sported seven PC-type ISA slots for expandability. This allowed the world of PC-compatible expansion

Figure 10.2

The TC-9 motherboard, which required the addition of a GIME chip and a CoCo 3 Super BASIC ROM to function. (Courtesy of Frank Hogg.)

Figure 10.3

The TC-9 as a complete system. (Courtesy of Frank Hogg.)

cards to come to an OS-9/68K system. It adopted a graphical environment called G-Windows, which was innately incompatible with the K-Windows offering of the other 68K systems.

In April 1991, Falsoft sponsored what would be its final RAINBOWfest in the place where it all started: the Hyatt Regency Woodfield in Schaumburg, Illinois. At that event, a roundtable was held between the CoCo 4 contenders. At the table were Lonnie Falk, Ed Gresick, Frank Hogg, and Paul Ward. Each made their case as to why their individual systems would be the right replacement for the CoCo 3.

Meanwhile, even as Tandy announced the demise of the CoCo 3, Falk and his staff at *THE RAINBOW* continued its routine planning and production of the magazine. It was fully expected that at some point, readership of a magazine devoted to a computer that was no longer selling would fall off. It was evident that the effect was real as issue after issue began shrinking in page count. Nevertheless, Falsoft continued to publish *THE RAINBOW* dutifully, even going to a newspaper-styled tabloid format starting with the March 1992 issue. Former Falsoft employee Greg Law recalls the change. "It [the tabloid format] came into being in order to save money so that the magazine could go on as long as it could. Lonnie really didn't want to give up *THE RAINBOW*."

Falk's apprehension to endorse any of the contenders for the title of "CoCo 4" was obvious in his editorials, but he did make one move that was seen by many in the community as favorable: He allowed reviews of the new machines to be published in the magazine and even went so far as to feature them on the cover, starting with Delmar's System IV on the September 1991 issue of *THE RAINBOW* (Figure 10.4). A cover spread and review of the Tomcat TC-9 came in

Figure 10.4

The September 1991 issue of *THE RAINBOW* featuring the CoCo 4 contender System IV.

the November 1991 issue (Figure 10.5), with the MM/1 taking the coveted cover slot in the December 1991 issue (Figure 10.6).

Despite the appearance of openness to the efforts of the 68K machines, the May 1992 issue of *THE RAINBOW* contained what many advocates of the systems thought to be a knife in the back. In that issue, Falk made a stunning admission in his "PRINT #-2" column: the 68K machines were not worthy of the title "Color Computer 4." Falk went as far as to tell his readers that he could not "in

Figure 10.5

The November 1991 issue of *THE RAINBOW* featuring the CoCo 4 contender Tomcat TC-9.

good conscience recommend that you step 'up' to any of the 68xxx computers."
He went on to state that the dearth of software for the new machines was something that would plague users, who would never fully enjoy the number of packages available for their Color Computers. Although his predictions would prove correct in the years that followed, his frank pronouncements stung many in the Color Computer community who felt that after Tandy turned its back, *THE RAINBOW* was turning its back, too.

Figure 10.6

The December 1991 issue of *THE RAINBOW* featuring the CoCo 4 contender MM/1.

In that same editorial, Falk admitted that he was putting advertising dollars at risk. The very purveyors of the new 68K machines would, expectedly, not advertise in a magazine whose stance was to discourage their purchase and use.

Further infuriation came with additional pronouncements by Falk in his July 1992 column, where he stated, "if you are going to look for another computer, I believe the obvious choice is not some poorly supported 68xxx-based machine, but an MS-DOS computer." Many saw that as Falk pushing subscribers of *THE RAINBOW* into his PC-based magazine, *PCM*. If it was not clear before, it was blatantly obvious to those who had invested in one of the CoCo 4 hopefuls that

THE RAINBOW would not be catering to them or their machine. What was not clear was what would become of *THE RAINBOW* itself.

When it did become clear, the news came suddenly and without warning as the May 1993 issue of *THE RAINBOW* began to arrive in mailboxes across the country (Figure 10.7). Readers were met with a somber four-word headline at the top of the front page: "The Time Has Come."

Figure 10.7

The touching front-page editorial from the final issue of *THE RAINBOW* magazine, May 1993.

Everyone who saw it knew immediately what it portended. This was going to be the final issue of their long-lived and well-loved Color Computer magazine. The time had indeed come.

The heading prominently featured the community's unofficial but beloved mascot, CoCo Cat, with his arms outstretched and head tilted, as if he too were resigned to what was happening to the CoCo community's longest running and largest monthly magazine.

Directly below the headline was a poignant poem by Stephen Foster:

Old Dog Tray's ever faithful;
Grief cannot drive him away;
He is gentle, he is kind.
I'll never, never find
A better friend than Old Dog Tray.

And thus began what is quite possibly the most touching obituary to a computer periodical ever written.

In this last column, Falk touted the resilience of a magazine that was delivered across the globe to numerous countries for nearly 13 years. As Falk proudly noted, *THE RAINBOW* was more than a mere magazine. It spawned marriages, brokered friendships, served military servicemen and women (even a CIA spy), and brought people together at 20 RAINBOWfest events throughout the country and through the many CoCo clubs and pen pal requests that were published.

By his own admission, Falk confessed that *THE RAINBOW* was losing money for Falsoft in the final two years of its life. It made no business sense to run a magazine out of one's own pocket, losing money all the while. Yet, out of a sense of sentimentality, Falk kept it going. It was no surprise to those who knew Lonnie.

The "Letters to the *THE RAINBOW*" section on page two belied the true gravity of the magazine's fate. Without any hint of finality, letters were published from eager readers wanting to learn more about this or that product, or get advice on which printer to use with their Color Computer. But aside from the standard fare of submitted articles, BASIC listings, and advertisers, the issue featured another harbinger of the end: letters of farewell from notables such as Radio Shack's Barry Thompson; columnists Marty Goodman, Steve Blyn, and Dale Puckett; and advertisers Chris Burke of Burke & Burke Systems, Tom Dykema of T&D Subscriptions, Tom Roginski of OWL Computer Services, and Howard Cohen of Cognitec, among other CoCo notables. It was a fitting tribute to the icon of the Color Computer community.

Near the very end of Falk's final column, he penned an excerpt from the third chapter of Ecclesiastes:

To everything there is a season, and a time to every purpose under the heaven:
A time to be born and a time to die; a time to plant and a time to pluck up that which is planted;
A time to weep . . .

In what would be his last words penned in *THE RAINBOW*, Falk admonished readers not to weep for the demise of their beloved magazine. "It forged a community of spirit. A commonness of purpose. A wonderful adventure. It was the instigator of lasting friendships. It touched us all, and we were all a part of it.

It was the greatest."

11

Hello Darling

The CoCo, along with the other 8-bit home computers of its era, faced a future relegated to increasing irrelevance as the mighty PC DOS (and later Windows) compatibles dominated both business and consumer markets. Survival meant turning inward and focusing on the core group of individuals and enthusiasts who continued to make up the Color Computer community. "Vintage" CoCo users from the 1980s along with newcomers began to band together to keep the flame lit, soldiering on to continue supporting the waning yet still fervent interest in the machine that they so loved.

This sense of community was maintained and strengthened by electronic information services such as Delphi and CompuServe, which continued to serve important roles into the early 1990s. The existing Color Computer and OS-9 forums on the respective services still received regular posts, but the cost of connecting was still prohibitive to most, and subscriber numbers would soon be steadily chipped away by a new fad called the "Internet." CoCo users rallied around this new medium to organize on both Usenet discussion groups, such as comp.sys.m6809, and other online mailing lists.

One such list, known simply as the "CoCo List," became an important area for enthusiasts to hang out during the 1990s. It was hosted by a list server at

Princeton University, which also managed mailing lists for other interests. It would later be merged into Usenet's newsgroup hierarchy as bit.listserv.coco.

Initially, the list was available to students and individuals who happened to have an e-mail account at a university or college, since Internet service at that time was relegated mostly to educational institutions that were linked together for research purposes. However, as the Internet expanded and grew into the 1990s, so did access to these mailing lists and, later, Web sites. With the burgeoning Internet, the CoCo community would find a place to not only survive but to eventually thrive. There were other possibilities for the future, however.

In April 1988, a Japanese magazine dedicated to Fujitsu FM-series personal computers, *Oh!FM*, published an article that would cause an explosion of enthusiasm within the CoCo world almost four years later. The article—written in Japanese, of course—revealed that Hitachi's 6309 microprocessor, a CMOS (complementary metal–oxide–semiconductor) variant of the Motorola 6809 found in computers like the CoCo, held a secret set of instructions and registers that were heretofore unknown (Figure 11.1).

Hitachi's 6309 was not a stranger to CoCo owners. For years, users were replacing their systems' 6809 microprocessors with the cooler running, more power efficient CPU. Additionally, Hitachi also manufactured a CMOS version of Motorola's 6821 PIA (peripheral interface adapter) known as the 6321. Companies advertised both of these chips in *THE RAINBOW* as a low-energy, 100% compatible alternative microprocessor to the 6809 and 6821. Aside from the necessary desoldering of the original parts and putting sockets in their place on the CoCo 3 motherboard, installation was easy due to 100% pin and package compatibility. The replacement process, although tedious, was fairly routine for those handy with a soldering iron.

The CoCo community's own Kevin Darling had experimented with Hitachi's 6309, noting that in certain cases, the processor seemed to behave slightly differently than the 6809 that he replaced in his Color Computer. For instance, upon encountering an illegal instruction (something that almost always precipitated a crash on the computer), the 6309 would skip subsequent op-codes differently than the 6809. Darling did not realize it at the time, but he had stumbled upon one of the hidden features of the remarkable processor.

The veil was lifted on February 23, 1992, when Hirotsugu Kakugawa posted a message titled, "A Memo on the Secret Features of the 6309," to the

Figure 11.1

The Hitachi 6309 microprocessor, an improved version of the Motorola 6809 containing additional registers and instructions.

comp.sys.m6809 Usenet newsgroup. Based on his Japanese-to-English translation of the April 1988 article in *Oh!FM*, Kakugawa brought to light the incredible features of the 6309 that had been hiding in plain sight for years. In addition to several additional 8- and 16-bit registers, the 6309 had some impressive characteristics: 16-bit hardware multiply and divide instructions, a 32-bit "super register," a special "native" mode where many instructions ran faster, and a super-fast "block move" instruction that could copy bytes in RAM at lightning speed.

The revelation of the 6309's hidden capabilities single-handedly breathed new life into the CoCo community at a time when despair was starting to set in, and only three short years after Tandy effectively abandoned the computer.

Tinkerers began to experiment with the processor's newly documented instructions, painstakingly verifying the information that Kakugawa posted. Within the year, Chris Burke had released a book, *The 6309 Book: Inside the 6309 Microprocessor*, on the Hitachi 6309 processor that explained each instruction. Burke followed that with a new software product called Powerboost, which patched Tandy's OS-9 Level Two to take advantage of the newer instruction set and faster operating mode, boosting performance of the operating system markedly.

The additional speed that could be achieved with the 6309 in native mode was attractive, but it came at a cost to OS-9 Level Two users. In this enhanced mode, the 6309 would stack its additional registers when an incident known as an "interrupt" occurred.* The addition of these new registers would throw off the entire register footprint that OS-9 was expecting, resulting in chaotic crashes. Burke's Powerboost ingeniously patched the interrupt service routine in the OS-9 kernel to expect this behavior and account for it, but it was not a complete rewrite of the operating system.

Meanwhile, a group of Canadian CoCo enthusiasts took a different approach. Curtis Boyle, Bill Nobel, and Wes Gale began the arduous task of disassembling the entire OS-9 Level Two operating system with the intent of baking in 6309 support directly into every component of the software. They spent months of grueling work converting the binary bits of 1s and 0s into readable assembly language source, painstakingly studied sections to decipher what was going on, then added comments to explain the functionality. Everything was open to disassembly and introspection, including the kernel, file managers, and drivers. It was a time-intensive reverse engineering effort that eventually resulted in the product known as NitrOS-9.

The use of the 6309's block move and bit mode instructions, combined with native mode operation and utilization of the extra registers, greatly improved the performance of OS-9 on the CoCo 3. The trio of programmers sold NitrOS-9 on 5.25" floppy disks as a total upgrade for Color Computer 3 owners who had moved over to the 6309. Sales through the mail and at CoCoFESTs were brisk, and several updates came in quick succession.

---

* The interrupt is a common event that causes a microprocessor to save its "state" or register set onto a memory stack, then temporarily execute code elsewhere.

Since 1983, Falsoft held at least two RAINBOWfests during the year; the dominant events were the Chicago RAINBOWfest in the spring and the Princeton RAINBOWfest, hosted in New Jersey around the fall. This changed in 1990, when Falsoft announced that the Princeton RAINBOWfest would not be held. Without a fall event that year, the Atlanta Computer Society of Atlanta, Georgia, decided to pull together to create the very first CoCo-related fest event in the Southeast, a tradition that continued for five more years, with the last Atlanta CoCoFEST being held on September 30 and October 1 of 1995. In the absence of Lonnie Falk's support, the CoCo community seemed resolved to pick up the slack.

The final Falsoft-sponsored event was the 1991 Chicago RAINBOWfest.

In 1992, the Chicago-area Glenside Color Computer Club worked with Color Computer vendor Dave Myers and his company, CoCo Pro!, to put on a "CoCoFEST" May 30–31, 1992, in Schaumburg, Illinois, the home to RAINBOWfests of years past. The event felt much like RAINBOWfests of prior years, with the notable exception that Falsoft was nowhere to be seen. The following year, Glenside would again put on a CoCoFEST, this time in the sleepy suburban town of Elgin, Illinois. Dubbed the "2nd Annual 'Last' Chicago CoCoFEST," the play on words was meant to convey a sense of impending finality to the event.

It was at this event in 1993 that Microware, the creator of OS-9, appeared for the first time (Figure 11.2). By this point, the company had long moved on from its 6809 roots and was an unlikely candidate to attend a CoCo-related event. However, an idea sparked by an employee gave the company a reason to attend. Having just upgraded its Intel-based version of OS-9 (known as OS-9000), Microware was left with dozens of boxes of the previous version of the software. In an effort to give Color Computer users a path to using OS-9 on the then-popular

Figure 11.2

Microware's booth at the May 1993 Chicago CoCoFEST.

80386 platform, Microware set up a booth and sold the normally $995 software for a much more reachable $350. CoCoFEST attendees snapped up the software.

Meanwhile, as 1993 approached, Terry Simons had an idea. Simons was a longtime Color Computer aficionado and founder of the Mid-Iowa & Country CoCo Club (known as MI&CC), a club based in the heartland of the United States, Des Moines, Iowa. MI&CC came into being sometime before 1992 and boasted a large membership that included members from all over the country (hence the addition of "& Country" to the club name, as Simons would often explain).

A U.S. Army veteran, Simons once logged 1,000 miles on a 3-speed bicycle over a two-and-a-half-year period while stationed in Germany. He also had the distinction of renting out a house to Ken Kaplan, founder of Microware Systems Corporation, when Kaplan was attending Drake University in Des Moines.

Simons lived in a quiet neighborhood in Des Moines along with his wife, Diane, and was also a small business owner, being the proprietor of Terry's Quality Concrete, where he performed residential and small commercial concrete work in the Des Moines area.

Although Simons owned and enjoyed other computers such as the Commodore Amiga, his first love was for the CoCo. He dabbled in BASIC programming, authoring a well-known home financial management package called *Home-Pac,* which he sold through his company, Computer Villa.

During Simons's leadership of MI&CC, monthly meetings were hosted at a local library in Des Moines, and members were provided a monthly "disk newsletter" that came in the mail on 5.25" disk. It was a unique distribution method for a newsletter: a subscriber would run the disk on a CoCo 3 and read various articles contributed by MI&CC members and others.

Few would argue that Simons's signature mannerism was his style of writing, both in newsletter articles and in his electronic messages: these messages always contained copious amounts of misspelled words, odd sentence structure, and liberal use of commas (e.g., "till then,,,,"). Often signing his posts with the handle "Terry g," he frequented online forums such as the CoCo List at Princeton, StGNet, and Delphi, where he was known as "MRUPGRADE."

In early 1993, Simons convinced the MI&CC membership to host a CoCoFEST right in Microware's backyard. As a result, the Middle America Fest took place March 27–28, 1993, on the outskirts of Des Moines, Iowa. This two-day CoCoFEST was the only one of its kind in the Des Moines area, having never taken place before and not to be repeated since.

The event took place at a hotel called The INN in Clive, Iowa, situated very close to Microware's headquarters. One of the more memorable events at the fest was the "Three Mugateers" corner booth where participants could take a digitized picture with Todd Earles and Mark Hawkins in-person, the Microware programmers who helped create the CoCo 3's internal ROM software and its version of OS-9 Level Two. Long-sleeved, gray T-shirts were sold for the fest, with a large yellow cat emblazoned upon the map of the continental United States.

There were concerns that the fest, which came only a few months before the 1993 Chicago fest, would either suffer in attendance or upstage the other event. Fortunately, both were well attended.

In spite of the many good things that Simons did for the CoCo community, both he and MI&CC became a source of controversy that permanently scarred both the man and the organization. The point of contention, which has become known as "The Orphanware Controversy," stemmed from the fact that MI&CC had collected a rather large catalog of free and no-longer-sold CoCo software, and offered its members access to the programs for a modest copying and shipping fee. The contents of this library became the subject of arguments between Simons and a number of CoCo software authors who discovered that copyrighted software was being offered in the library for a small fee, all without any agreement or permission of the original authors.*

In 1994, messages critical of MI&CC's favorable stance on orphanware began appearing on the Princeton CoCo List. In the months that followed, the discussion ebbed and flowed, culminating with Simons's active participation on the list in early 1996. It was then that Simons argued openly in favor of MI&CC's library, going so far as to state that if the copyright holder objected to the inclusion of the software in the library, only then would it be pulled. Many notable CoCo software authors were involved in the discussion, including Chris Burke, Joel Hegberg, Alan DeKok, Eric Crichlow, and Chet Simpson.†

Software authors were enraged that anyone would place the burden of ownership and claim on the software author. The position of the software authors was clear: respect the copyright and obtain permission before putting it in the club library. The culmination of the controversy came in early 1996 when Eric Crichlow, who worked for Microware at the time and lived in the Des Moines area, met face-to-face with Simons at a bookstore in West Des Moines, Iowa. Crichlow contended that Simons and MI&CC were blatantly violating copyright. Simons asserted that he was doing nothing wrong by offering software that was no longer being sold to CoCo owners who still wanted to use it. No minds were changed at that meeting, and Crichlow was determined to bring his concerns to the community.

Later that year, Crichlow attended the 1996 Chicago CoCoFest, bringing with him the now-infamous "WANTED Poster," which proclaimed Simons as a danger to the CoCo community and advocating that people not join MI&CC in order to take a stance on orphanware. The fest booklet even contained a compelling piece written by Crichlow, stating the particular danger of allowing orphanware to become an acceptable part of the CoCo community.

---

* This type of debate rages to this day within nearly every classic computer and videogame community. On the one hand, the letter of the law regarding copyright is clear, on the other hand, if a product is no longer available for sale anywhere, there are few alternatives for a particular platform's enthusiast to otherwise gain access to the software, particularly for emulation.

† Incidentally, Chet Simpson was coined "The Saint" by Terry Simons himself, in an attempt to cast Simpson as a hypocrite for his alleged participation in the cracking of copy-protected CoCo software some years earlier. Simpson gladly used the moniker in his posts, and even developed a CoCo game that he named after the gag: *Gold Runner II: Return of the Saint*.

From that point on, MI&CC seemed to have lost its luster as the CoCo community largely shunned Simons's efforts. In spite of some demand for MI&CC's orphanware library, a number of members of MI&CC either did not renew their subscriptions or canceled outright. Other organizations within the CoCo community, such as the Glenside Color Computer Club, made their positions clear, stating that they were not in support of the actions of MI&CC and Simons. In the end, both MI&CC and Simons faded into obscurity. Simons formally closed the club in 1998, sold a great deal of his personal CoCo collection, and moved on to the PC world.

On May 18, 2000, Simons passed away from complications of throat cancer at the age of 58. He is buried at the Des Moines Masonic Cemetery in Des Moines, Iowa.

* * *

THE RAINBOW's decline and eventual demise in 1993 portended an uncertain future for professional CoCo publications that were dependent on mass readership. This decline allowed more homegrown periodicals to again pick up the slack. One such magazine, CoCo Clipboard, came by way of Ted and Darlene Paul. As a former Radio Shack employee, Ted Paul knew the CoCo well and set out to create a magazine that would serve the interests of the community. It did so for several years before ceasing publication before 1991.

Another well-known periodical that was embraced by Color Computer owners was World of 68 Micros. Its owner, Frank Swygert, a true CoCo enthusiast, had publishing experience related to his vintage car hobby and decided to extend this venture to his favorite home computer. When THE RAINBOW announced it would cease publication, Swygert contacted Falk hoping to fulfill remaining subscriptions with his own. When Falk turned him down, Swygert asked to purchase Falsoft's mailing list, but the price was way outside the budget for the fledgling magazine maker.

"What perplexed me most at the time was that Mr. Falk had always told us loyal subscribers that he had the best interest of CoCo people at heart first and foremost," says Swygert. "I understand he wanted THE RAINBOW to die with dignity instead of letting some unknown person assume operation on a smaller scale. But I didn't understand an obviously high price for the mailing list [for a hobbyist on a limited budget]. I even asked about paying Falsoft to send a one-time mailing out instead of giving me the mailing list, but that would have been more trouble than it was worth to him and nearly as costly to me."

In August 1993, Swygert printed his first issue with 150 prepaid subscriptions. By the end of the second year, the operation had doubled to 300 subscribers. Articles focused not only on the Color Computer, but the 68000-based computers that had earlier vied for the title of the "CoCo 4," including the MM/1 and TC-70—the very computers that Falk had earlier indicated THE RAINBOW would not support.

Swygert's World of 68 Micros topped out at just under 400 subscribers before starting a slow decline, with his last issue making its appearance in the spring of

1997. At that point, Swygert had competing interests and could no longer devote the time that he needed to the magazine. Not willing to see the magazine disappear, he instead found someone who was willing to take on the continuing tasks of publishing and mailing. Unfortunately, that individual failed to meet the obligations, and the magazine ceased publication.

*　*　*

The CoCo community had its own spiritual advisor in Brother Jeremy, an Anglican monk of the Order of the Community of Saint Joseph the Worker (Figure 11.3). An avid CoCo user, Brother Jeremy could be seen attending CoCoFESTs in full habit. As a fan of OS-9, he became obsessed with the unreleased OS-9 Level Two Upgrade that had been worked on by Kevin Darling and others. In a most determined way, this man of the cloth made it his mission to bring the upgrade out of the darkness and into the light for all to see and use.

For years on end, Brother Jeremy continually asked Darling to release the OS-9 Level Two Upgrade, incomplete as it might be, just so that others could enjoy it. Darling, who was presumably hamstrung by legal dictums from Tandy and Microware that prohibited him from even speaking about the project, let alone give it away, remained mum for years, even long after Tandy discontinued the CoCo 3.

Not to be dissuaded, Brother Jeremy created what has to be the most peculiar request in modern computing history: a song by a monk, imploring a computer programmer to release the code for an operating system upgrade for a defunct computer. And no less, put into words and recorded! The song, addressed directly to Darling himself and aptly titled *Hello Darling*, provides a window into the plaintive calls of the CoCo community to release the unfinished product. A few

Figure 11.3

The CoCo community's monk and spiritual advisor, Brother Jeremy, CSJW, who tirelessly petitioned for the release of the secretive OS-9 Level Two Upgrade. (Courtesy of Allen C. Huffman.)

of the original OS-9 Level Two Upgrade developers were named specifically in this humorous homage:

> Hello Darling, it's been a long time,
> Since you worked on Level Two OS-9.
> But I'm begging you now,
> Please hear my plea.
> Kevin Darling, set the upgrade free.
> Well, I spoke to Kent Meyers today.
> Mark Griffith, he gave his OK ...
> Bill Dickhaus said that the contract was dead,
> It expired back in '93
> So I'm leaving it all up to you.
> And you know what I hope you will do.
> So I'm begging you now,
> Please hear my plea.
> Kevin Darling, set the upgrade free.
> Last night, as I sat by the screen,
> I must have been having a dream.
> For what did I see,
> on the old CRT,
> but the words "OS-9 Version Three."
> So I'm leaving it all up to you.
> And you know what I hope you will do.
> So I'm begging you now,
> Please hear my plea.
> Kevin Darling, set the upgrade free.
> So I'm begging you now,
> Please hear my plea.
> Kevin Darling, set the upgrade free.

Perhaps it was the song or perhaps it was plain old "monk persistence," but eventually Kevin Darling relented and provided Brother Jeremy with distribution disks of the OS-9 Level Two Upgrade.

* * *

Something was bothering Eric Crichlow.

The former CoCo developer who authored ShellMate, a handy utility program for OS-9 Level Two, had moved onto the burgeoning "CoCo 4" scene with a game he wrote: *Gold Runner 2000*, which ran on the 68000-based MM/1 from Interactive Media Systems. The computer was superior in graphics and processing to the fading CoCo 3.

What stymied Crichlow was the less than optimal performance of the game that he wrote for the 68000-based CoCo 4 contender. True, *Gold Runner 2000* was written in the C language instead of the more primordial but higher performance assembly language of the 68000 processor, but he expected the 16-bit computer to provide much better performance.

He certainly expected it to still outperform the CoCo 3.

Crichlow was one of a number of CoCo developers who had moved on to the new 68000 machines, resigned to the fact that the CoCo 3's capabilities, however impressive back in 1986, were no longer up for the challenge. It was simply not possible to provide equal gameplay to his new MM/1, he thought. *Gold Runner 2000*, being a side scrolling game, required either massive CPU heavy lifting to copy large amounts of pixels, or special "screen scrolling hardware" that allowed the graphics chip to cleverly shift the screen's contents around.

On top of this, each level in *Gold Runner 2000* took up 576K of memory, and it played a continuously looping background soundtrack that changed for each level. The sound files for the soundtrack to achieve this feat were relatively huge. In addition to that, the MM/1's sprites used 24 colors, and the CoCo 3 was only capable of displaying 16 simultaneous colors, though in a fairly decent 320 × 192 resolution.

With all of the perceived power stacked in favor of the MM/1, Crichlow felt that it was impossible for a CoCo 3 to come up to that level, and voiced his opinion to his longtime friend and CoCo enthusiast Chet Simpson. Simpson countered, asserting that not only could he do such a game for the CoCo 3, but he could make it better than the MM/1 version that Crichlow wrote (Figure 11.4).

And so a challenge was born.

Crichlow, confident that he was correct, told Simpson that if he were to attempt such a feat, he would have to use the same graphics, level, and sound assets as the MM/1, although in order to give the CoCo 3 a fighting chance, he conceded to some modifications to this requirement. Simpson was allowed to redevelop the

Figure 11.4

Eric Crichlow and Chet Simpson selling their CoCo and MM/1 software at the 1994 Atlanta CoCoFEST. (Courtesy of Allen C. Huffman.)

CoCo: The Colorful History of Tandy's Underdog Computer

graphics, using 16 colors—though he had to still push about as many pixels per screen update as the MM/1 version. And he was allowed to resample the music to attempt a reduction in the size, so long as the sound quality was not noticeably degraded.

The game had to also include a static score bar at the bottom of the screen. This stipulation alone could present performance problems with the full-screen animation on similarly classed computers. Finally, the game had to be completed and available for sale at the 1997 Chicago CoCoFEST. Crichlow was betting that due to the insurmountable technical challenges, Simpson would not finish the project.

Simpson went to work. Weeks before the deadline, he had taken a new job in Los Angeles, which required that he spend time packing for the move. Just prior to the move, he showed Crichlow the current state of the game—it was mostly done and was shaping up nicely, but he had not yet been able to get the score bar at the bottom of the screen without removing the background music, due to lack of memory.

As the 1997 Chicago CoCoFEST came around, Crichlow, who did not attend that year, learned that Simpson's game was not available for sale at the fest. Fairly and squarely, Crichlow declared himself the undisputed winner of the challenge. But eventually, Simpson did complete the game, and to Crichlow's surprise, it ran faster and better on the venerable CoCo 3 than it did on the perceptibly more powerful MM/1. "To be fair, he [Simpson] did an absolutely amazing job on the game, and I believe he eventually got everything required into it," Crichlow conceded years later.

Simpson's feat restored Crichlow's faith in the CoCo 3's potential. Even though the market for such games was not nearly the size in 1997 as it might have been 10 years earlier, Simpson had proved that the CoCo 3 still had its chops, and that a determined developer could push the limits and make a game that still sold. Crichlow's earlier doubts on the CoCo 3's capabilities, raised from the utterance of statements made at the Princeton RAINBOWfest in October of 1986, were erased.

The event that brought so much doubt to Crichlow was the CoCo 3 Roundtable Discussion, held just months after the CoCo 3 was released in late 1986. The roundtable consisted of prominent people in the development of the Color Computer 3. Barry Thompson and Mark Siegel from Tandy were two of the participants, as were the very highly regarded CoCo game programmers Dale Lear and Steve Bjork. *THE RAINBOW*'s Lonnie Falk moderated the discussion that evening, with the audience volleying a number of questions to the panel.

One particular question was directed to Bjork, who was asked what he envisioned being able to do with the CoCo 3's advanced graphics. Bjork answered by asserting that he had been attending user group meetings for the Apple Macintosh, Atari ST, and Commodore Amiga, and studying the latest and greatest games available on those platforms.

Bjork summed up his view of the software offerings of the CoCo 3's high-end competitors at that point: "Not anywhere close to what we're working on. In other words, basically a lot of stuff that you've seen for the ST and the Amiga as far as its capabilities, as far as its graphics, we will be able to exploit them with the Color Computer 3 and we will do them.

"I will not comment at this time exactly what we're going to do, but I have to say I am quite pleased with what I'm doing, and it probably makes *Zaxxon* look like a *Pong* game."

Bjork was believed to have been referencing the game that he brought to the Color Computer while working at Datasoft. *Zaxxon* was a well-known and often ported Sega arcade game from the early 1980s, and Bjork's adaptation to the CoCo was seen as one of the more impressive versions. Given Bjork's history and status in the community, it did not seem unreasonable to grant some serious weight to his assessment.

Crichlow indicated that he waited with baited breath for amazing games to appear—games that would indeed make users of those higher-end systems jealous. Later, Bjork did release several CoCo 3 titles, including *Warp Fighter 3-D*, a space shooter that made optional use of red-cyan 3D glasses; *Mine Rescue*, a platformer that bore a striking resemblance to *Super Pitfall*; and *Z-89*, a greatly enhanced but unofficial port of *Zaxxon*. Yet Crichlow asserts that while some of those were decent games, no reasonable person could argue that any of them came anywhere near living up to the bar that Bjork had set so high for himself with his October 1986 statement. "Knowing that he [Bjork] had been arguably the best CoCo programmer of the CoCo 1 and 2 era, and not having seen anyone else come out with much better than what he had done for the CoCo 3, I surmised that perhaps the problem wasn't Steve, perhaps the problem was that the CoCo 3 just wasn't as capable as he had made it out to be."

But after seeing what Chet Simpson was able to do, Crichlow rethought his view of the CoCo gaming great. "I came to realize that Steve's mouth had just written checks that his talent couldn't cash."

Crichlow continues, "It's really a shame that Chet, and others, didn't do more with the CoCo 3 earlier on. I would really like to have seen what its true potential was."[*]

Despite the regrets, the CoCo 3's full gaming potential was at least hinted at under one of the most prolific, yet undercelebrated, game programmers of the day, Glen Dahlgren. Dahlgren started out writing games for the CoCo 2, including the *Hall of the King* trilogy, *White Fire of Eternity*, and *Kung-Fu Dude,* the latter of which detected when it was running on a CoCo 3, allowing it to both run faster and play better sound effects. Most of his work was sold through Prickly Pear Software, a prolific publisher of Color Computer games. With the advent of the CoCo 3, however, the extent of Dahlgren's software development talents became fully realized. He designed, wrote, and sold a a number of quality, action-packed, full-color games for the platform through his company,

---

[*] As is usually the case, it often takes years for programmers to learn all of a platform's technical nuances and really push it to its limits. Some platforms are fortunate to survive long enough to have the majority of their potential realized within their commercial lifetimes, while others need years longer. As you will read in the next chapter, the CoCo 3 did get to fulfill more of its potential starting in the late 1990s.

Sundog Systems, which would often occupy the coveted, but expensive, inside front covers of *THE RAINBOW* magazine's advertising space. The impressive array of titles his company published included arcade clones, *Sinistaar* and *The Contras,* as well as the original puzzle game, *Photon,* each of which nicely taxed the CoCo 3's graphics and sound capabilities. *Sinistaar* itself required a full 512K CoCo 3, and took three disks just to load!

* * *

As PC compatibles rapidly gained more speed, memory, and performance capabilities throughout the 1990s, it reached the point where they were able to properly emulate older platforms. Jeff Vavasour had an idea to do just that with the CoCo.

Growing up with his parents' TRS-80 Model I, Vavasour was already familiar with Tandy's line of computers, including the Color Computer, but it was not until 1993 that he actually purchased one for $20 at an antique store. He resolved to write an emulator on an MS-DOS-based 386 PC, and the result was the first freely available, downloadable CoCo 2 emulator, *TRS-80 Colour Computer 2 Emulator.*

The extent to which Vavasour performed the emulation was impressive. Joysticks on the PC would take on the personality of CoCo joysticks, allowing virtually all of the games that could run on the Color Computer to run on the emulator. It was a strange sight indeed to see a PC running a full green 32 × 16 screen showing the familiar sign-on text from Tandy's BASIC ROM, not to mention actually running OS-9.

Not content with stopping there, Vavasour would later tackle the CoCo 3 (*TRS-80 Colour Computer 2 Emulator*), a somewhat more difficult machine to emulate due to its enhanced speed, graphics, and memory requirements. Ever the prolific programmer, he would also write emulators for the TRS-80 Model I, III, and 4.

Beyond the extent of Vavasour's work, others would bring their own emulators to the scene. These included David Keil's *CoCo Emulator,* Steve Bjork's *VCC,* and the open source *MESS* (Multiple Emulator Super System) project, which emulates not only the CoCo, but a large variety of other computer and video-game platforms.

Emulators have arguably extended interest in the CoCo over the years by giving those without the original hardware a means of more easily experimenting with the BASIC environment. Although by no means a "real" Color Computer (and looked down by some as a faux means of using a CoCo), a PC-based emulation setup can provide a rewarding, if not exactly pure, experience.

Whatever the moral position one may have on emulation and its use of downloadable software files of questionable legal standing, there is no disagreement that the Color Computer's true heart and soul is the original hardware inside the elegant battleship gray or off-white case emblazoned with the TRS-80 or Tandy logos. For the CoCo's biggest fans, nothing beats using an original.

12

CoCo Forever

The relentless march of technological progress litters the road with the carcasses of computer chassis from bygone eras. This progress has long left behind the CoCo in terms of being a modern tool with significant utility and function. Yet, despite that nondebatable tenet, the Color Computer is still tinkered with and hacked on, and actively traded on auction sites like eBay. Software and hardware are still made for the CoCo even today and sold to enthusiasts who are eager to push the machine to its absolute limits.

Just how many Color Computers were manufactured and sold since it was first introduced has the aura of a state secret. The exact number has never been revealed, and when asked, no Tandy ex-employee would divulge the total. There are hints, however.

A clue can be found on page 190 in the February 1987 issue of *THE RAINBOW*, which has this quote: "The speaker is Barry Thompson, the man who buys and sells Color Computers in the Tandy Towers at Fort Worth. 'If you're looking for opportunity, look no further. The potential installed base is in the millions. In fact, the installed base of the Coco 2 is already in the millions.'"

In the May 1993 issue of *THE RAINBOW*, some four years after the CoCo 3 was canceled, Thompson wrote: "I have written purchase orders for more CoCos than anyone else in the world. I stopped counting them when it passed a million."

If the serial numbers of Color Computers are any indication, the number is well past 2 million. Yet former Tandy officials who were in the know, like Bernie Appel, John Roach, Mark Yamagata, and others, would not give an answer when interviewed some 20-plus years after the last Color Computer was sold. When Mark Siegel was asked directly how many Color Computers were sold, he would not give a specific number, only to say, tantalizingly, "It was a lot... a lot more than people might realize."

As of this writing, nearly 23 years have passed since it ceased to be manufactured, yet the CoCo still has not died in quite the way some had predicted. Instead, it arguably went into a state of light hibernation for a time. Fans still carried the banner and held onto their systems.

Today there is a renaissance of interest in home computers and videogames from the 1970s and 1980s. Retrocomputing is in vogue, and the CoCo, as well as its contemporaries, are actively bought and sold many places online and off. There is seemingly no shortage of Color Computer hardware or peripherals for the person who is willing to pay the price for them.

Events like the CoCoFEST, which have been the glue that has held the CoCo community together for over two decades, continue to be hosted and well attended. The Chicago-area's Glenside Color Computer Club has continued to host yearly CoCoFESTs nonstop since its first event in June 1992. Though much smaller in attendance than the glory days of Falsoft's RAINBOWfests, people still gather from all over the United States (and occasionally from other countries) for a weekend of reminiscing and nostalgia. Vendors sell hardware and software, and there is always a new project or two in development that is being shown.

Years after its official demise, amazing new software continues to be developed for the Color Computer as well. As an illustration of that progress, an impressive videogame conversion project was completed by longtime Australian programmer Nick Marentes in 1997. A videogame developer who had sold multiple titles to Tandy for both the TRS-80 and CoCo back in the day, Marentes designed a clone of the classic arcade game *Pac-Man* for the CoCo 3, incorporating surprisingly accurate audiovisuals (Figure 12.1). Despite the lack of the more vertically oriented 244 × 288 resolution of the arcade game, Marentes managed to squeeze all of the key gameplay elements into the CoCo 3's much wider, and shorter, 320 × 192 graphics screen. A few years later, in 1999, Marentes would go one step further and release an amazing first-person shooter, *Gate Crasher* (Figure 12.2), inspired by the *Gloom* 3D engine that John "Sock Master" Kowalski previously released as a demo in 1996 to prove that a *Doom*-style game could be possible on the CoCo 3.

In 2007, Kowalski performed an equally impressive feat of software engineering, porting the original Z-80-based version of the arcade game, *Donkey Kong*, including all of its graphics and sound assets, to the CoCo 3 (Figure 12.3). Many have hailed it as the best and most accurate rendition of the arcade classic on any 8-bit home computer. In order to obtain the best and closest possible rendition of the original, Kowalski painstakingly mapped the Z-80 assembly code over to 6809 assembly language by hand. This was no easy feat, but the resulting work

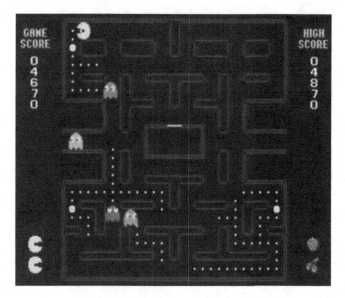

Figure 12.1

A screenshot from Nick Marentes's impressive *Pac-Man* conversion.

Figure 12.2

A screenshot from Nick Marentes's incredible first-person shooter *Gate Crasher*.

shone through. All four levels (Barrels, Factory, Elevators, and Rivets) are in the game, with play virtually indistinguishable from the original.

Even the older Sierra-published graphical adventure games, such as *King's Quest: Quest for the Crown*, *King's Quest II: Romancing the Throne*, *Space Quest: The Sarien Encounter*, *Manhunter: New York*, and others, which never originally found their way onto the CoCo 3 platform, are now very much playable today on the system. It was discovered that the Sierra AGI (Adventure Game Interpreter)

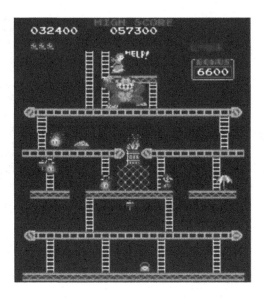

Figure 12.3

A screenshot from John "Sock Master" Kowalski's arcade faithful port of *Donkey Kong*.

game engine that was ported to the CoCo 3 for the *King's Quest III* and *Leisure Suit Larry* games worked equally well with other Sierra games of the era. Thus was born a whole new set of playable, interactive adventures. Ironically, this would have also been easily achievable back in 1988, but for unknown reasons, neither Tandy nor Sierra bothered to bring these other packages to the CoCo 3.*

The OS-9 operating system, which was the cornerstone of the Color Computer for much of its life, has been resurrected in "The NitrOS-9 Project," an open source collaborative effort between members of today's CoCo community. Using cross assemblers and other tools running on modern personal computers, NitrOS-9 bootable images can be quickly assembled and deployed in a matter of seconds. Support exists for all Color Computers, as well as the Dragon series.

More recently, the availability of inexpensive processors and chips have led some to find ingenious ways of keeping the CoCo experience alive and thriving. Instead of emulators written in languages that are compiled for other (namely Intel) processors, the use of densely populated programmable logic devices, known as field-programmable gate arrays, or FPGAs, have opened new avenues for breathing life into older systems. Hardware engineer Gary Becker and software wizard John Kent are two such individuals who have successfully created 6809 and Color Computer 3 FPGA implementations, respectively. With evaluation boards around $150, it is possible to have a "bare metal" Color Computer in hardware with programmable logic.

---

* In fact, despite contacting the likes of Sierra cofounder Ken Williams and other notables from the company at the time, there is little to no remembrance of the work done on the CoCo 3, including who exactly did the original AGI conversion!

Such approaches have fueled debate in not only the CoCo community but in the larger retrocomputing field about the validity of emulation or simulation versus the original system. Torn between resurrecting old hardware using new techniques, or attempting to locate, refurbish, and use older hardware, different camps have different ideas on what is acceptable. Many continue to use emulators, whereas some have drawn the line at the FPGA implementation as a fair equivalency of an actual Color Computer. Others believe that a CoCo is not truly a CoCo unless it has the original case, parts, and keyboard, with no substitutions.

The CoCo community has also found clever solutions to the ever-dwindling supply of antiquated floppy drives, controllers, and diskettes. DriveWire is a combination of a protocol and a connectivity system that allows CoCo users to access disk software from the convenience of their PC by using a simple serial cable.

Longtime Color Computer supporter and hardware guru Mark Marlette continues to do his part to keep the CoCo community alive, providing innovative hardware and software through his company Cloud-9, out of Delano, Minnesota.* Marlette has designed, manufactured, and marketed an SCSI controller, an IDE/CompactFlash interface, memory boards, keyboard interfaces, and more.

The aforementioned Australian CoCo enthusiast Nick Marentes continues his quest to solve the mystery of the "256 color mode" in the CoCo 3's GIME chip. Such a mode, thought to have existed, would bring even more vivid color usage and techniques to games and other programs. Evidence suggests that such a mode was at least in the early planning stages of the CoCo 3 and probably made it into the early prototype board.

The mad monthly rush to the mailbox for the latest issue of THE RAINBOW magazine has been long quieted. Yet where print has stopped, the World Wide Web has taken over. Today, Web sites containing CoCo information in the form of articles and photos can be found all over the Internet, as do online forums and the popular "Malted Media CoCo List" run by longtime CoCo legend Dennis Bathory-Kitsz.

* * *

Before the demise of THE RAINBOW in 1993, Lonnie Falk confided in Jim Reed an ambition that he had: running for political office. Reed, whose disagreements with his old boss had become the stuff of legend, instinctively admonished Falk. "No way in hell will you win." Undaunted, Falk took Reed's dismissal as a challenge, and ran for mayor of the town of Prospect anyway . . . and he won.

Not only did Falk win one term, he ran for several more, totaling 13 years of service to his home community as its mayor. Even ex-THE RAINBOW columnist and former Falsoft employee Ed Ellers came on board under Falk's leadership.

Falk was still mayor of Prospect when he passed away on June 9, 2006, at the age of 63. Jim Reed is still in the Louisville, Kentucky, area.

* * *

---

* Full disclosure: The author Boisy G. Pitre, along with Aaron Wolfe, is also a key part of Cloud-9.

If the CoCo community had a true teacher, it was William Barden, Jr. A prolific author in his own right, Barden worked with Tandy's publications manager Dave Gunzel, as well as Radio Shack book buyer Leon Lutz, to provide great books for Color Computer users. Along with other authors like Radio Shack employee and ham operator Harry Helms and Forrest Mims III, respectively, Barden's writings would reach beyond the Color Computer and into electronics and computers in general.

Today, Barden continues his writing endeavors in a different genre, but still looks fondly upon the CoCo. "It was a fantastic machine. The CoCo was a compact computer with the ability to easily interface to the outside world." Even today, Barden says that he has never found a computer that could be used by an average person quite like the CoCo could. "It was simple: hook it up to your TV, turn it on, and in mere seconds you can start doing equations or type simple programs."

*   *   *

Microware arguably extended both the life and the usefulness of the Color Computer, as the machine continued well into the 1990s and beyond. Aside from bringing OS-9 to the Color Computer, the company continued to adapt OS-9 to new processors and markets, including interactive television set-top boxes and cell phones. After going public with an IPO in 1996, Microware was purchased in 2001 by Radisys of Hillsboro, Oregon. More recently, OS-9 assets were acquired by a resurrected Microware LP consisting of a trio of companies that have committed to continue selling and supporting the operating system for modern processors.

Microware's original founders are no longer associated with the company, having gone down their own paths. Ken Kaplan, who left Microware in 2001, now heads Cellencor, an Iowa-based company focusing on microwave systems for industry and agriculture. Larry Crane and Robert Doggett have opted for less visible roles, continuing to work in the software industry.

Two of the CoCo 3's famous "Three Mugateers" also continue to work in the software arena. Mark Hawkins owns his own consulting firm and resides in Iowa; while Tim Harris, who left Microware to work for Microsoft, is now in the business of developing mobile applications and resides in the state of Washington. Regrettably, Todd Earles passed away in 2007.

*   *   *

Upon retiring from Tandy in 1999, John Roach headed various initiatives, including the John V. Roach Honors College at his alma mater, Texas Christian University. He still holds the Color Computer in high regard, and even opined on an opportunity that Tandy may have missed. "If you want to wonder, and I wonder about it, but I can't go back to the time and know what all the thinking was, is whether we could have done something with OS-9, or more with OS-9, because it clearly drew a lot of supporters."

Green Thumb and VIDEOTEX's engineering father Jerry Heep was still working at RadioShack in August 2012 when we were extended and accepted an invitation to visit its headquarters and sift through some of the remaining Tandy archives that Heep had at his Fort Worth office. One month after our visit, Heep left the company after 37 years of service and is now enjoying retirement.

Tandy's Mark Siegel and Barry Thompson spent a great deal of time both advocating for and defending the continued existence of the Color Computer. Their advocacy and efforts kept the CoCo in Radio Shack catalogs and stores for 10 years, along with a plethora of hardware add-ons and software packages. Although Thompson passed on years ago, Siegel continues to work in the electronics design industry and still resides in Texas.

In May 1993, Tandy's computer business was sold to AST. Most of the engineers were laid off, but Dale Chatham was one of the few that was asked to stay onboard. Instead of remaining, Chatham took a job with OASIX, an ASIC design company in Boulder, Colorado. After OASIX was sold to Texas Instruments, Chatham ended his career as an electrical engineer. He is now retired and residing in Pilot Point, Texas.

*   *   *

A special trip has been in the works for three weeks now, and the day has finally arrived for me [Boisy G. Pitre] to leave. It's a beautiful and sunny Friday afternoon in early April when I start the seven-hour drive from Prairie Ronde, Louisiana, to Austin, Texas. Early along the route, I'm surrounded by acres of flooded, crawfish-filled rice fields that border Louisiana Highway 13, but in an hour, the landscape changes to flat grassland along the I-10 corridor and into Texas. I stop in the sleepy town of Huffman, just outside of Houston, to visit my cousin Mitch Pitre and his family, where I rest for the night.

The next morning I wake up early and head for the rolling hills of Highway 290. A detour onto a winding country road takes me through the countryside where longhorn cattle and ranches dot the landscape. The view is spectacular as my iPhone guides me on through the towns of Brenham and Bastrop. A few hours later, I am nearing the south end of Austin, where I take a brief jog up I-35 and into the thick brush and stony hill country.

The sole purpose of the trip is to meet the engineer behind the Color Computer 2 and 3, and the father of the GIME chip. The day has turned cloudy, and a cool breeze permeates the air. As I approach the area where John lives, I carefully pace my driving in order to arrive exactly at the agreed upon time of 11:00 A.M. Driving slowly through the neighborhood, I eventually locate John's house. As I pull up slowly into the driveway, a thin man with glasses and a Gandalfish white beard comes out of the house to greet me. This is John Prickett.

He invites me into his home, and we walk into the living room. There on his couch sit two intriguing pieces: a Deluxe Color Computer that he picked up at a Radio Shack internal auction for $5, and something that looks amazingly like a CoCo except for a 3.5" floppy drive jutting out of the top. The keyboard appears as though it was pulled from a CoCo 3, and the badge along the top left

Figure 12.4

The CoCo 4 mockup is the only one of its kind in the world.

reads: "Tandy Color Computer 4." John goes on to explain that the mockup was designed by Tandy industrial design engineer Bernard "Bernie" Grae as a concept for a CoCo 4 (Figure 12.4). The case is hollow and empty.

I begin to ask John questions: his foray into computers, his formal education, and how he began working at Tandy. For the next three hours, I sat next to the man whose engineering talent and skills were used to create the computers that made Tandy a lot of money, as well as endeared a legion of faithful fans. Never far from his coffee machine, John told story after mesmerizing story with the skill of an expert campfire raconteur, taking well-paced pauses to sip from his coffee cup.

At some point during our conversation, he casually mentions that he has serial #0000001 of the Color Computer 2. Without hesitation, I ask to see it. He obliges, taking me on a short walk out to a storage building in his backyard, and opens the door. A stream of boxes fills the building from bottom to top, but he quickly spots the CoCo 2 box. "Ah, there it is!" he exclaims, and we head back to the living room, where John carefully opens the box.

Inside is the computer, still being held in place with the familiar two-piece Styrofoam guards. He gently removes the computer from the box, which contains the BASIC programming manual and a TRS-80 Color Computer 2 Service Manual. Turning the computer over, we look at the bottom label and indeed, it is the first CoCo 2 to come off the assembly line (Figure 12.5 and Figure 12.6). "Kenji Nishakawa [Tandy's CoCo factory manager] personally handed this to me," explains John.

The appearance of the computer brings a barrage of questions from me that John happily answers. His memory is sharp and reliable; when he is unsure of something, he unambiguously states so. While our discussion continues, I notice a large framed photo of what appears to be a Color Computer 2 sitting on a desk with red, green, and blue orbs hanging suspended above. John notices my interest

Figure 12.5

The top of John Prickett's CoCo 2 Serial #0000001.

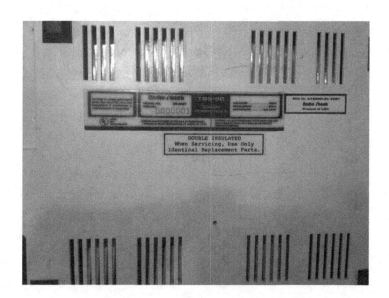

Figure 12.6

The bottom of John Prickett's CoCo 2 Serial #0000001.

Figure 12.7

John Prickett stands next to a framed picture of the CoCo 2 that he created.

in the picture, walks over and grabs it off the wall, then proceeds to show it to me (Figure 12.7).

John looks at the picture, and says, "I rescued this from an office at Tandy. They were going to throw it away." John, like the rest of the CoCo community, clearly knows when something is worth saving and treasuring.

Index

Nelson Software Systems' Super "Color" Writer II, 38

New Pilot Magazine, 60

New York Stock Exchange, 3

Nimrod, 5

Nintendo, 48

NitrOS-9, 157, 172

NTSC standard; *See* National Television System Committee standard

NWS; *See* National Weather Service

O

Odyssey Home Entertainment System, 10

Online forums, 159

Open source collaborative effort, 172

Orphanware Controversy, The, 160

OS-9, 80
    creator of, 158
    forums, 155
    Intel-based version of, 158
    operating system, 45, 59, 92

OWL–WARE, 68

OXO, 5

P

Pac-Man clone, 170, 171

Panzers East!, 44

PCB; *See* Printed circuit board

PC DOS compatibles, 155

PCM, 59, 60

Peripheral interface adapter, 156

Personal computers; *See* Tandy gets personal

Personal Finance, 39

PFA; *See* Professional Farmers of America

Phantom Graph, 121

Photon, 167

Phreakers, 12

Pinball Wizard, 70

Planetfall, 44

Planting the seed, 19–28
    bidding process, 20
    CompuServe, 24, 28
    connectivity to the outside world, 22
    custom services, 26
    direct connect modem, 22
    expectations, 20
    farmer's "electronic almanac," 20
    Farm Information Retrieval System, 19
    government contracts, 20
    graphics capability, 21
    Green Thumb, 19
    hobbyist Bulletin Board Systems, 25
    joint design, 21
    "Ma Bell" telephone system, 20
    MicroNET, 24, 27
    Motorola-designed 6847 Video Display Generator, 21
    National Weather Service, 19
    newspaper articles, 23
    printed circuit board, 20
    QWERTY keyboard, 21, 25
    radio frequency interference, 22
    selling agreement, 26
    software distribution service, 24
    TV output, 25
    unlikely genesis into color computing, 19
    U.S. Department of Agriculture, 19
    VIDEOTEX, 24
        external features, 25
        pricing, 27
        TRS-80 Terminal, 26

Plateau of the Past, 45

PLATO computer instruction mainframe system, 7

Pong, 11, 166

Pooyan, 44

Popeye, 43

Popular Computing magazine, 15, 30, 31, 36

Popular Electronics magazine, 7, 9

"Porting" the operating system, 94

Powerboost, 157

Preannouncement of upcoming product, 132

Printed circuit board (PCB), 20

Product personality, 46

Professional Farmers of America (PFA), 25–26

Profit margins; *See* Double trouble

Prototype boards, 108

Pseudocolor issue, 74

Q

Quasar Commander, 39, 40, 41

Quick Printer II, 34

QWERTY keyboard, 21, 25

R

Raaka-Tu, 45
Radio-Electronics magazine, 71
Radio frequency (RF) device emissions, 17
Radio frequency interference (RFI), 22, 32
Radio Shack, 1, 122
    accuracy of catalog, 132
    catalog, 1, 2, 8, 131
    employee, 161
    franchise owner, 146
    internal auction, 175
    stores, CoCo 3's debut in, 120
RAINBOW, THE (magazine), 53–68
    ad hoc submissions, 63
    advertising, 61, 62, 67
    "Barden's Buffer," 63
    brainstorming, 65
    cartoon strips, 63
    CoCo Cat, 63, 64
    contact with readership, 61
    copyediting, 63
    creation of cover illustration, 57
    debut artwork, 56
    demographic of the era's pioneering
        computer owners, 53
    editorial operations, 60
    escape route, 60
    Falsoft, 55, 56, 60
    first formal edition of, 55
    first RAINBOWfest, 59
    fledgling magazine, potential of, 56
    golden era for Falsoft, 59
    Hurricane Camille, 54
    improvements to magazine, 56
    influence of famed illustrators, 57
    in-house staff, 64
    interesting moments, 58
    magazine pitch, 57
    memorable RAINBOWfest, 68
    modest start-up, 53
    NASA's first moon landing, 54
    OS-9 operating system, 59
    OWL–WARE, 68
    payment milestones, 59
    philosophy, 65
    phone calls from subscribers, 61
    Q&A column, 63
    software listings, 62
    submission from readers, 62
    techie material, 62
    technical review, 62
    trust, 63
    United Press International, 53
RAINBOWfest, 59, 68
Random-number generator, 41
Raster Management System (RMS), 102
Raster Memory Interface, 102
Rescue on Fractalus!, 122
Reverse engineering, 157
Reverse video, 38
RF device emissions; *See* Radio frequency
    device emissions
RFI; *See* Radio frequency interference
RF modulator, 80
RMS; *See* Raster Management System
Robot Odyssey I, 44
Rocket, The, 139
ROM chips, 91
RS-DOS, 43

S

Sailor Man, 43
S-100 bus, 9
SCADA, 85
Scorecard, 60
Screen scrolling hardware, 164
SCSI-based hard drive package, 137
Sega arcade game, 166
Shanghai, 132
Sierra Adventure Game Interpreter, 171
Sierra On-Line, 122
Sierra-published graphical adventure
    games, 171
Silicorn Valley, 89–99
    ad placement, 91
    BASIC language, 91
    C compiler system, 92
    CDC-6600 computer, 90
    Children's Computer Workshop, 96
    CoCo community, mission of
        education to, 99
    company perk, 96
    daunting task 92
    Disney titles, 96
    engineer-turned-president, 94
    factory farms, 89
    FLEX operating system, 94

Printed in the United States
by Baker & Taylor Publisher Services